COUNTDOWN TO GCSE

GEOGRAPHY

Richard Daugherty

Lecturer in Education
University College of Swansea

MACMILLAN
EDUCATION

ACKNOWLEDGEMENTS

The author and publishers wish to acknowledge the following sources: Michael Mayes of Nicholas Chamberlaine School, Bedworth; also London and East Anglian Group; Midland Examining Group; Northern Examining Association; Southern Examining Group; Welsh Joint Education Committee; Northern Ireland Schools Examinations Council.

The publishers have made every effort to trace the copyright holders, but if they have inadvertently overlooked any, they will be pleased to make the necessary arrangements at the first opportunity.

First published 1987
Reprinted 1988

Published by
MACMILLAN EDUCATION LTD
Houndmills, Basingstoke, Hampshire RG21 2XS
and London
Companies and representatives
throughout the world

Printed in Great Britain by
Cox & Wyman Ltd, Reading

Designed and illustrated by
Plum Design, Southampton

British Library Cataloguing in Publication Data
Daugherty, Richard
Geography. (Countdown to GCSE)
1. Geography   Study and teaching
(Secondary)   Great Britain   2. General
Certificate of Secondary Education
I. Title   II. Series
910'.76   G76.5.G7
ISBN 0-333-41344-X

CONTENTS

Countdown to GCSE: Geography

1	Introduction	1
2	About GCSE: What's new?	7
3	Aims and objectives	10
4	What's on the syllabus?	15
5	Which type of assessment?	25
6	Understanding coursework	32
7	Writing coursework	45
8	Understanding the examination papers	61
9	Answering examination questions	90
10	Preparing for the examination	117

SECTION I

Introduction

Your grade for GCSE Geography will depend partly on how well you can answer examination questions that stretch over several pages like this:

1 **Employment and economic influence**

   (a) **Study Map 1 below.**

Map 1: Unemployment in the United Kingdom – 1970 and 1982

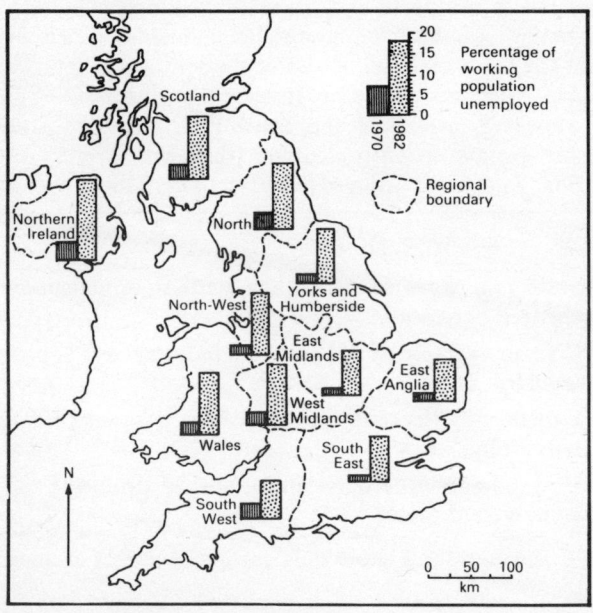

Answer the following questions.

(i) **Which region had the highest unemployment percentage in 1982?**

(ii) Complete the table below.

Changes in unemployment in the West Midlands

unemployed in 1982	. . . %	
unemployed in 1970	3%	
Thus unemployment in 1982 was	. . . %	higher than 1970

[3 marks]

(b) **Read Extract 2 below.**

> *Extract 2*
> *Jobless total grows*
> With its partners in the Common Market, the UK has seen a
> sharp rise in unemployment over the last twelve years. People
> of working age still find it more difficult to get a job when
> they live in the north and west of the country. There, unem-
> ployment still hits the traditional industries much harder
> than the 'high tech' industries of the south and east. And, of
> course, Northern Ireland has its special problems.
>
> However, people in the South-East have seen a much
> bigger *growth* in unemployment than almost every other
> region. And this in spite of the attractions of the South-East.

Answer the following questions.

(i) State *two* ways in which the pattern of unemployment
changed between 1970 and 1982. *[2 marks]*

(ii) Give an example of a 'traditional industry' and a 'high tech'
industry. *[2 marks]*

(iii) Explain why the traditional industries have suffered particularly badly. *[3 marks]*

(iv) How have Northern Ireland's 'special problems' affected
employment there? *[4 marks]*

(c) **Map 1 and Extract 2 show only regional patterns of unemployment.**

Describe *two* local variations which are lost in the regional
figures. Give a *named* example for each. *[6 marks]*

(d) **Study Diagram 3 showing percentage employment in each of
the main sectors of industry in 14 countries in 1981.**

Diagram 3

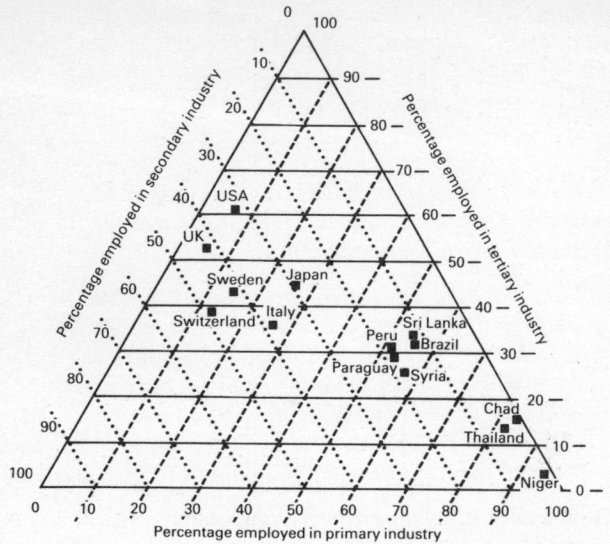

(i) **Complete the table below showing employment percentages in Sri Lanka.**

Primary sector	. . . %
Secondary sector	12%
Tertiary sector	. . . %

[2 marks]

(ii) **Explain what is meant by the terms 'primary industry' and 'tertiary industry', giving *one* example in each case.**

[6 marks]

(iii) **It may be said that primary industry is of great importance in the least developed countries, while highly developed countries rely on tertiary industry. To what extent is this statement borne out by the facts shown in Diagram 3 and knowledge of any studies undertaken?** *[12 marks]*

(SEG B)

This question is typical in many ways of the questions you can expect to be asked in a GCSE Geography examination. It is mainly about one topic, unemployment, but extends to distinctions between different types of industries and different levels of 'development'. The scale of study ranges from patterns within a UK region through the UK as a whole to the world scale. What you have to do varies from extracting specific geographical information to discussing the wider issue of industry in least developed countries. To answer the question you will have to read maps, graphs and text, remember information from your studies and explain situations you have not previously encountered.

Your first reactions?

'Easy. I could answer a lot of it without following the course.'
Perhaps so, because general knowledge and general skills are undoubtedly helpful. But beware of over-confidence.

'I knew I should have chosen History instead. I could never answer a question like that.'
Don't write yourself off. Before being faced with such a question, you would have studied the topic during your GCSE course and practised the skills you need for answering it. Maybe you *can* answer the question if you take things step-by-step, understanding what is required and learning to make the most of what you know and can do.

'I doubt if I will have time to read the questions, still less to answer them.'
It's true that geography examination papers are getting longer. Gone are the days of a single sheet of paper with short, blunt questions like:
What is the capital of France? or
Write an essay on world copper production.
Question papers are longer so that they can cover the range of knowledge and variety of skills which geography calls for.

Unfortunately some of your geographical abilities will not be assessed in the examination paper. You may be brilliant at using a map to find your way through an unfamiliar area but the examination will not credit you directly for being able to do that. You may be good at explaining things orally but geography examinations judge your ability to write rather than to talk. Don't allow those abilities to be devalued just because GCSE examinations do not recognize them. But do learn to identify what *is* being tested by the examination papers. If you have understood the topics and developed the skills covered during your course, sections 8 and 9 of this book will help you demonstrate your abilities to best effect in the examination.

Being able to answer examination questions will not be enough in itself to assure you of success in GCSE. Some of the work you do during your course will be marked by your teacher and will count towards your final grade. You might undertake a fieldwork project like this:

1 *Aims*

To investigate the impact of public use upon a leisure area. For example:

How is the area altered to provide leisure facilities?

What effect do people using the leisure facilities have on the area?

2 *Fieldwork*

Assess the quality of the leisure landscape using a landscape assessment scheme.

Record instructions to the public.

Record damage to, or neglect of, the facilities.

3 *Processing*

Map of landscape attractiveness.

Distribution map of instructions to the public.

Trace overlay of damage to leisure area.

Assess the effectiveness of the instructions.

Compare leisure areas.

4 *Conclusions*

Answering the question(s) posed at the beginning of the study. For example:

extent of impact, reasons for impact, similarities/differences between leisure areas, what can be done to reduce adverse effects.

[Based on an example in: Jennifer Frew, *Geography Fieldwork*, Macmillan]

Your reactions?

'What exactly is a leisure area? How do I assess landscape quality?'
If this were *your* coursework, you would have the advice and help of your teacher. A project such as this could be a class exercise, with only the writing up of results left to each student to do in his/her own way.

'When would I find the time to do all that?'
Coursework is built into the GCSE course. Much of the work is likely to be done during geography lessons though you may find there is additional fieldwork or writing up to do.

'Studying leisure areas sounds boring.'
Leisure may not be an aspect of geography which interests you. This example was given only to illustrate the *type* of work involved in undertaking a GCSE project. There are innumerable possible topics you could study in your coursework. The choice of topic is for your teacher or, where the whole project is done individually rather than as a class exercise, yourself. Only three conditions limit that choice. The study must be geographical in nature, linked to a syllabus topic and involve

you in some fieldwork. Sections 6 and 7 of this book explain more fully what coursework requires of you and how to tackle it.

The unemployment question and the leisure areas study are examples of the type of work GCSE Geography involves. The next four sections of the book explain the background to the work you will do, dealing with questions such as:

What's new about GCSE?
What are the aims of GCSE Geography?
What will I find in the syllabus?
What type of examination and coursework can I expect?

The final section (10) offers some suggestions for organizing your revision in the final weeks of the course.

There should be no mystery about what you will have to do to succeed in GCSE Geography. Of course, the examination papers are secret until you are given the signal to open them. Of course, the markings of your answers will be done in secret to be fair to all candidates. But, whether setting the questions or marking the answers, examiners are following guidelines about what can be asked and what to give marks for. Similarly, when setting or marking coursework tasks, your teachers are following guidelines so that, though the work will differ from school to school and candidate to candidate, everyone is judged on a comparable basis.

The aim of this book is to help you understand what a GCSE Geography course and examination will expect of you. Ultimately your grade will depend on the geographical abilities you can demonstrate when doing coursework or answering examination questions. The book does not set out to teach those abilities but to help you make the best use of them whether you are striding across the moors in pursuit of your project or seated in the exam room puzzling over question 3 (b) (ii).

SECTION 2

About GCSE: What's new?

Examinations taken mainly by sixteen-year-olds have been with us for a long time. So what is different about the General Certificate of Secondary Education (GCSE)?

SYLLABUSES

The syllabuses for GCSE, unlike previous examination syllabuses, conform to national guidelines – the 'National Criteria'. So, will the course you follow be much the same as that for every other GCSE Geography candidate in England, Wales and Northern Ireland? Perhaps surprisingly, no. As will be evident in the next two sections, the national guidelines are loose enough to allow for a variety of different syllabuses. Each of the groups planning a GCSE Geography syllabus has taken account of current thinking to design a course for the late 1980s and early 1990s. In some cases this has meant modifying a pre-GCSE course; in others, an entirely new course has been designed for GCSE.

COURSEWORK

The way in which your work will be assessed must also conform to the National Criteria. So, can you expect the same type of coursework and exam questions as will be set for other GCSE Geography candidates? Again, not necessarily.

In one respect, study outside the classroom, GCSE will ensure that what has been common practice in good geography teaching will feature in every course. 'Fieldwork' is a compulsory part of the assessment to be made by teachers during GCSE Geography courses.

Such coursework assessments, undertaken by teacher(s), are to figure in all GCSE courses. The minimum requirements are that coursework must account for at least 20% of the total marks and must include some fieldwork. Beyond that, the type and amount of coursework varies markedly from one syllabus to the next (page 33). But you can be sure that some of what you do before you get to the examination room will have been assessed by your own teacher(s) and credited to you. In this way your GCSE course will allow you to demonstrate abilities which you cannot show when answering examination papers.

EXAMINATION PAPERS

When it comes to the final examination part of the assessment, the designers of each syllabus have chosen the methods they think are appropriate. They decide which types of questions to use, how long the papers should be and whether there should be common papers for all candidates or separate 'easier' and 'more difficult' papers to choose from. They have to accept that it is possible to assess only a small part of what you know and can do and to carry out that assessment in a limited, imperfect way. But they have used the best available methods to try to ensure that the grade you will be awarded is a fair and accurate indication of your geographical abilities.

GRADES

The GCSE scale of grades from A to G replaces the grades A to E awarded for GCE O-level and 1 to 5 for CSE. How will it be decided which of the grades A to G your work in the GCSE examination and coursework merits? In a few years' time it may be that there will be clear targets you have to meet – a specified level of performance – to be awarded a given grade. For the moment, however, which grade you will be awarded will be decided by the familiar method of adding up all your marks. Those achieving the highest marks will be given As, the next most successful group Bs and so on down the scale.

FINDING OUT MORE

This book explains the features which are common to all or most GCSE Geography syllabuses. To find out more about the syllabus on which your course is based and the way you will be assessed, you will need a copy of that syllabus. Your teacher may be able to supply one. Indeed, for a syllabus which is designed wholly or partly by the school/college (as some are), that is the only possible source of the information.

But you are more likely to be following a course based on one of the sixteen GCSE Geography syllabuses prepared by an examining group. In that case, the school or college budget may not extend to producing multiple copies of a document which may run to a hundred pages in length. Find out which examining group and which of its syllabuses you are dealing with and write off for a copy to the address given below.

Ask for specimen (or past) examination papers as well as the syllabus. You may have to pay, but for a full picture of your GCSE Geography course you can't do without the syllabus and the examination papers.

Most of the examining groups have prepared several GCSE Geography syllabuses in order to offer schools and colleges courses with different geographical emphases and varying methods of assessment. For those groups which have more than one Geography syllabus, an index letter is used to identify which is which. Where an extract from a syllabus or an exam question is quoted in this book, its source is indicated by the abbreviated group title and the syllabus index letter. For example, the question on unemployment at the beginning of the book (pages 1–3) was taken from the Southern Examining Group's B syllabus, hence the label SEG B.

Addresses

There are four examining groups in England, each with several constituent examining boards. Separate examining boards exist for Wales and Northern Ireland. The addresses of the boards responsible for GCSE Geography syllabuses are as follows:

London and East Anglian Group, The Lindens, Lexden Road, Colchester, Essex CO3 3RL

Midland Examining Group, Robins Wood House, Robins Wood Road, Aspley, Nottingham NG8 3NR

Northern Examining Association (NEA, A and B), 12 Harter Street, Manchester M1 6HL

Northern Examining Association (NEA, C and D), 31-33 Springfield Avenue, Harrogate HG1 2HW

Southern Examining Group (SEG A), 23-29 Marsh Street, Bristol BS1 4BP

Southern Examining Group (SEG B), Beloe House, 2-4 Mount Ephraim Road, Royal Tunbridge Wells, Kent TN1 1EU

Welsh Joint Education Committee, 245 Western Avenue, Cardiff CF5 2YX

Northern Ireland Schools Examinations Council, Beechill House, 42 Beechill Road, Belfast BT8 4RS

SECTION 3
Aims and objectives

AIMS

Statements of aims indicate in a general way what it is hoped you will gain from the course. Some aims refer to the nature of geography:

(a) **To develop a sense of place and an understanding of relative location.**
(b) **To develop an awareness of the characteristics and distribution of a selection of contrasting physical and human environments.**
(c) **To develop an understanding of some of the processes affecting the development of environments.**
(d) **To promote an understanding of the spatial effects of the ways in which people interact with each other and with their environments.**

(NEA A, B, C)

Others are concerned with the application of geography to circumstances which we meet in our everyday lives:

To use important ideas and skills drawn on in geography to classify and interpret such everyday experiences as discerning order in landscape.
To consider environmental, political and social issues which have a geographical dimension at local, regional and world scales.

(MEG D)

To foster better understanding of different communities and cultures both within our own society and elsewhere in the world.

(LEAG B)

To promote an awareness of the way in which individuals can participate in the process of change or conservation at various scales.

(Avery Hill)

Aims also refer to the kinds of learning you will be engaged in during your course.

The development of a range of general skills which relate to the acquisition and communication of knowledge.

10

The development of a range of enquiry skills through practical work, including investigations in the field, associated with observation, collection, representation, analysis, interpretation and use of data, including maps and photographs.

(SEG A)

To develop problem-solving skills.
To develop communication skills.
To develop the ability to generalize from the particular and to formulate and test hypotheses.
To be able to make, on the basis of sound evidence, judgements and decisions on economic, environmental, political and social issues and problems.

(MEG C)

The examination papers and coursework will test your proficiency in such skills but there are also less definite, more personal, gains which course designers and teachers hope for. For example:

To develop in candidates a sense of values through an understanding of behavioural, social, economic and political forces which influence decision-making.

(MEG A)

To provide experiences which will stimulate and encourage pupils' interest in and awareness of the environment and encourage positive participatory involvement in creating the world of tomorrow.

(Avery Hill)

For many GCSE syllabuses the general statements of aims are very similar, even though the details of the courses differ. In some cases, however, the way a syllabus's aims are expressed is the first clue to the distinctive character of that syllabus. For example, one syllabus puts particular stress on the importance of being aware of people's attitudes and values when studying geography:

Using appropriate geographical knowledge, to encourage and enable candidates to:
(i) appreciate the significance of people's values and attitudes on their perception of the world and their actions in it;
(ii) explore values and attitudes in relation to
conservation and change of the physical and human environment;
decisions about the management of the human and physical environment;
spatial and social inequalities;
the contrasting opportunities and constraints facing people living in different places under different physical and human conditions;

(iii) **make a significant contribution to the development of values and attitudes conducive to the elimination of inequalities, including those determined by racism, social structure and sexism.**

(Avery Hill)

OBJECTIVES

Aims reveal the thinking of the people who designed the course. Objectives are more detailed statements of the abilities which will be tested. One way to find out what is expected of you is to look at the examination questions and coursework tasks, but the objectives do give some indication of what the examiners will test you on.

Much of the examination is concerned with finding out whether you have understood the topics studied during the course. You will need to show that you understand landscapes and patterns, how they change and why that happens. To:

demonstrate an understanding of the wide range of processes contributing to the development of
(i) **the various environments and their associated landscapes;**
(ii) **spatial patterns and interactions which are important within these environments.**

(NISEC)

You will also need to show that you can apply what you understand to examples which you will be presented with for the first time in the examination. So you should be able to:

demonstrate an understanding of the geographical ideas, concepts, generalizations and principles specified in the syllabus and an ability to apply these in a variety of physical, economic, environmental, political and social contexts.

(SEG A)

Demonstrating that you understand something may well involve showing that you can both describe and explain it. You will be expected to:

demonstrate a capacity to describe, and to seek explanations for, the interrelationships in the environment of physical and economic activities.
(WJEC)

You may also be expected to put forward an opinion, with supporting argument, on a current issue. To:

make reasoned appraisals of economic, environmental, political and social issues in geography, and values and attitudes in decision-making.
(WJEC)

Studying geography involves developing and practising skills of gathering, presenting and interpreting information. Some skills, such as map use, are distinctive to geography while other skills are common to many subjects. The range of skills is wide. For example, one syllabus expects you to be able:

to select and use a variety of techniques appropriate to a geographical enquiry, including investigation in the field, and, in particular,
(a) **to use basic techniques for obtaining, observing, recording, representing, analysing, classifying and interpreting data;**
(b) **to use a range of source materials, including maps at a variety of scales, photographs and simple statistical data;**
(c) **to depict information in a simple map and diagrammatic form e.g., drawing line graphs, pie charts, divided bar graphs, triangular graphs, radial charts, flow charts, relief sketch-sections, sketch maps;**
(d) **to demonstrate an ability to select, use and communicate information and conclusions effectively.**

(LEAG B)

To back up your understanding and skills you must have a basis of factual knowledge of the topics covered in the course. This knowledge will sometimes be tested directly, at other times it will be tested together with the testing of your understanding. It may be only one item in a long list of objectives but it is fundamental to the examination that you will be able to:

recall specific facts and demonstrate an adequate level of locational knowledge relating to syllabus content.

(WJEC)

Balance of objectives

One important message not contained in a list of assessment objectives is the *relative* importance of each of the objectives in contributing to the total assessment of your geographical abilities. How far, for example, will a good memory, which enables you to recall facts with ease, take you? The answer to that kind of query can be found in a table in your syllabus which summarizes how much credit will be given for each broad category of ability in each component of the assessment. For the MEG D syllabus, the balance of objectives assessed in each component of the examination is as follows:

Assessment objectives	Written paper	School-based assessment		Weighting %
	Paper 1 or Paper 2	3 course-work units	Geographical enquiry/ies	
Knowledge recall	30	15	5	25.0
Understanding	30	15	10	27.5
Skills (including practical skills)	20	15	20	27.5
Values	20	15	5	20.0
Total marks	100	60	40	200 marks 100%

(MEG D)

From that it will be seen that if the abilities tested are in the proportions intended, even a perfect memory will serve to gain you no more than the 25% overall weighting allocated to 'knowledge'.

Such weightings are notoriously difficult both for examiners to carry through in setting questions and for teachers to maintain when assessing work done during the course. For one thing, what for you may be a test of understanding could in effect be a test of recall for another candidate who happens to have encountered the same question before. Or, what the examiner intends as a test of your map-reading skills could become for you a problem of how to explain what you have detected on a map.

But though a gap remains between the examiner's intentions and the actual abilities you have to demonstrate, that table of objectives and the examination components are the best general guides to the way in which you should use your time preparing for the examination. The point will be taken up again in section 10 but what should be clear already is that it will not be sufficient to do well in only one respect, whether it be remembering facts, using maps or applying what you understand to a 'new' situation. Success in your GCSE Geography course will depend on demonstrating your ability in terms of each of the objectives mentioned in your syllabus.

SECTION 4

What's on the syllabus?

All GCSE syllabuses must conform to the National Criteria but there are significant differences between the syllabuses. To find out exactly what *your* syllabus says, get hold of a copy from your teacher or direct from the examining group and study it. What this section can do for you is:

(i) tell you in broad terms what you can expect;

(ii) explain how to make sense of the syllabus your course is based on.

You may be inclined to let your teacher do the thinking about what the examining group stipulates should be studied. But it can help you to work more effectively if you understand the framework of requirements around which your teacher has designed the course you are following.

Examination syllabuses have come a long way from the days when naming a topic such as 'Africa' or 'Glaciation' was considered sufficient information for student and teacher to be clear about what to study. Your GCSE syllabus describes at some length the subject matter on which you will be examined. Though several pages long, the description of the syllabus content still leaves scope for the individual teacher or student to decide how to deepen, extend and illustrate the topics listed. That scope may offer a welcome freedom to you and your teacher to study the course in your way but it can also be a source of uncertainty. In what depth should this topic be dealt with? How far does that topic extend? Which examples, and how many of them, should be looked at? No amount of careful reading of the syllabus, even when backed by studying past examination questions, will give unambiguous answers to questions such as these.

Yet, while you cannot expect the boundaries of what you should study to be drawn as boldly as the political boundaries in your atlas, much can be learned from an initial reading of the syllabus and by referring back to it while you are studying the course. Even if the outer limits of a topic remain uncertain, the syllabus gives strong signals about what is central to your course. If you are to avoid the disappointment

15

and frustration of examination questions unfamiliar to you, it is essential to pick up and interpret those signals.

THEMES

There are some themes which are so central to studying geography that you can expect them to be included in your course:

Population - distribution, structure, growth, migration.

Settlement - patterns and processes, especially in urban areas.

Economic activity - location of, effects of, changing patterns; especially industrial and agricultural activities.

Natural environments - processes shaping them; environmental impact on human activities.

Other themes sometimes included are:

Contrasts in economic development

Resources

Energy

Transport and trade

Leisure and tourism

The titles of the syllabus sections or modules are the first clues to the kind of geography you will be studying. Compare these two lists of themes:

The syllabus is set out in five themes as follows:

Theme A	**Physical environment**
Theme B	**Population and settlements**
Theme C	**Agriculture**
Theme D	**Transport**
Theme E	**Industry**

(MEG B)

The subject content is contained in the four modules:

Module 1	**Employment and economic influence**
Module 2	**Leisure, recreation and tourism**
Module 3	**The divided world**
Module 4	**Environmental management**

(SEG B)

16

FINDING OUT MORE

It may seem, as you are working through your course, that you are accumulating a depressingly large amount of information on a lengthening list of topics. It will help your study, and especially your preparation for the examination papers, if you recognize that behind the many topics and examples covered are three key elements in everything you do:

1 *Skills* which you will learn and practise during the course.
2 *Geographical ideas*, an understanding of which you will develop by studying specific topics.
3 *Locations*, the places and areas you will learn about.

At every stage of your course it is helpful to focus on these key elements. Having them in mind before the course enables you to see where you will be going. It can also help during the course if you are unclear as to what is important in a topic you are studying. It is absolutely vital at the revision stage (section 10) if you are not to be submerged under the pile of notes you have made as you go along.

Skills

The skills to be mastered are usually listed for the course as a whole rather than for each topic. The list of skills could be something like those shown in the table on the following page.

Faced with such a list, you may wonder whether you can cope with all the skills referred to. But remember, it is a programme of skills for the whole, usually two-year, course. Your teacher will include the teaching and practice of each skill at appropriate points over that period.

Ideas

Whether the syllabus calls them *key ideas, concepts* or *generalizations*, they have the same function – highlighting the main ideas which are at the heart of understanding a theme. Superficially, your course may consist of a series of case studies such as these suggestions for population geography:

1 **The journey to work.**
Commuting and its effects including 'rush hour' problems upon a large United Kingdom urban area or conurbation.

Intellectual skills	Related techniques
Reference skills – ability to make use of a variety of sources for obtaining information.	– data collection through fieldwork. – data collection from various media-type/slide book/journal/film, etc.
Communication skills – ability to present information in a clear and appropriate way.	– *Transformation of data: into graphs (line, histogram, pie, radial, scatter); maps (sketch, choropleth, isopleth, topological)* – landscape sketching.
Interpretative skills – ability to give meaning to data.	– *Interpretation of data: graphs (line, histogram, pie, radial, scatter); maps (sketch, choropleth, topological, OS – at scales 1:50,000, 1:25,000, 1:10,000); photographs (vertical air, oblique, ground level); text, diagrams; numerical data.* – *Simple network analysis.* – *Analysis of documentary evidence e.g., old maps, photos, advertisements*, questionnaires, etc. – *Analysis of atlases.*
Evaluative skills – ability to consider evidence and form a conclusion.	– role play exercises/games. – *Decision-making exercises.* – discussion.
Conceptualizing skills – ability to organize information to form a concept or generalization.	– discussion. – research and investigation. – prediction.
Hypothesizing skills – ability to formulate hypotheses and to test and reformulate on the basis of evidence.	– classifying. – *Application of understandings to new situations.*

The candidate's ability to use the techniques in *bold* in the table above may be assessed in the written papers. All the techniques listed are appropriate to the candidates Individual Study, particularly those which are not assessed in the written papers.

(WJEC)

2　The effects of the growth of public transport systems and of the use of the private car upon a large urban area.

3　Studies of population pyramids typical of a less developed country and of a more developed country to gain an understanding of the terms 'birth rate' and 'death rate' and of the expectations of life at birth.

4　A study of a densely populated river valley in a less developed country to gain an understanding of the term 'over-population'.

5　A study of the causes of high population density and low population density, including high population growth rate and low population growth rate in two contrasting rural areas and two contrasting urban areas.

6　A study of population changes in the United Kingdom and inter-regional migration from areas of economic decline to areas of economic growth.

7　A study of the depopulation of an upland area.

(NEA A)

But the case studies are not an end in themselves, rather a means to the end of understanding key ideas, which for that section of that syllabus are shown in this way:

Population geography

Key ideas

(a) (i) Population is distributed unevenly and the density of population varies.
　　(ii) Distribution and density are affected by a variety of physical, political and socio-economic considerations.

(b) Population growth and structure vary world-wide.

(c) (i) Some parts of the world are over-populated, others under-populated.
　　(ii) Over-population and under-population have important influences on social and economic development.

(d) There are both short-term and long-term, and permanent and temporary movements of population.

Small-scale	Regional national scale	International global scale
Short- and longer-term changes in: (i) the distribution of population; (ii) population structures. The causes and effects of daily population movements.	The causes and effects of variations in the structure and growth of population in: (i) a less developed country; (ii) a more developed country. 'Push-pull' factors and population movements. Migration between rural and urban areas. Problems caused by population growth; concepts over- and under-population and their relationships to standards of living.	The factors affecting the variations in the distribution, density and growth of world population.

(NEA A)

When it comes to setting the exam questions, it is your understanding of ideas such as these which the examiner will be testing. A question would not expect you to remember all about, say, the pattern and problems of commuting in the London area. It would require you to show, with reference to such an example, that you understand 'the causes and effects of daily population movements'.

Locations

'If it's geography it must be about some*where*.' But the practice of stipulating places or areas which must be studied is less and less common. Most syllabuses leave the choice of places and areas for study to the teacher. The exam questions in most cases do not therefore expect knowledge of any particular places but they will require you to use what you know about the places you have studied.

You could find yourself studying a theme through examples drawn from selected regions or from anywhere in the world. For instance, the key ideas for a module in one syllabus are set out thus:

Heading	Key idea	Commentary
Systems	1 The world's water supply is contained within a closed system – the water cycle. 2 Individual drainage basins form dynamic open systems with inputs, throughputs and outputs.	The components of the water cycle include precipitation, run-off and evapotranspiration. The system structure of drainage basins permits the analysis of water movement in terms of stores, flows and losses.
Classification	3 Landforms and landscapes associated with rivers and river valleys and coastal zones may be classified according to their characteristic appearances and to the processes which form them.	Processes include weathering, mass movement, erosion, transport and deposition. The landforms created may also be influenced by rock type and structure.
Spatial patterns	4 The natural availability of water varies from place to place. 5 There is often a mismatch between supply and demand.	Physical factors include rainfall patterns and river systems. Areas of high water supply often occur in remote, thinly populated and/or mountainous regions.
Dynamics	6 The supply and distribution of water can change as a result of natural fluctuations and human intervention in the water cycle. 7 Competing demands for water can create conflicts, both environmental and economic. 8 Natural and man-made changes to the coastal zone may create opportunities and/or problems.	Changes may include floods, their causes, effects and control; drought and the process of desertification; water management schemes; pollution. These may be local (e.g,. flooding a valley for a reservoir), national (e.g., water transfer schemes) or international (e.g., drainage basin boundary conflicts). Opportunities include resort development; new port sites; land reclamation. Problems include cliff recession; silting of ports; the need for sea defences; pressures of tourism.

(LEAG A)

21

For that module some suggested examples are given, *either* with reference only to the UK and Brazil *or* drawn from anywhere in the world:

The suggestions below for teaching outlines for the syllabus modules are merely illustrations of possible approaches and are in NO SENSE PRESCRIPTIVE.

Core module – Landscapes and water

A A teaching outline using only the United Kingdom and Brazil	Syllabus key ideas illustrated	B A teaching idea using examples drawn from a variety of locations	Syllabus key ideas illustrated
Water cycle – general	1	A study of the water cycle: global scale	1
UK precipitation distribution	4	Rainfall regimes and patterns in Great Britain and Nigeria	4
Sample Pennine Drainage basin study – as a system		A study of a drainage basin in Great Britain	2
– landforms and processes, human effects	2, 3, 6, 7	Contrasting river valleys in different parts of Great Britain (e.g., Wharfedale, Langdale, Cuckmere or Thames)	3
Mole and Thames, N Wales – floods: causes, effects and control; transfer schemes; demand for water; uses of reservoirs, conflicts: Dinorwic and Lake Vyrnwy.	4, 5, 6	Flooding in the river valley: the Severn Valley; Mississippi; Nile; Niger	6
Water supply and demand in UK	5, 7	Water supply and demand in (i) Great Britain (ii) West Africa	5
Amazon Basin } precipitation NE Brazil } – distribution } and contrasts	2, 4, 7	Water distribution in Great Britain (e.g., transfer schemes, barrages)	5, 6
Amazon: source to mouth – processes and landforms, regimes	2, 3	Multi-purpose water schemes: Tennessee Valley Authority; Murray-Darling	6, 7
NE Brazil: drought, effects, responses	4, 6	Drought and desertification in West Africa; The Sahel	4, 6, 7
Itaipu scheme: Brazil/Paraguay cooperation, environmental impact	6, 7	Competing demands for water as a scarce resource in Great Britain (e.g., Kielder, Welsh water), in SW USA, Mexico, in the Indus Valley	7
Coastal study of the Isle of } Purbeck Coastal study of Norfolk } rock types, processes, landforms, environmental issues.	2, 3, 8	Coastal Studies: contrasts in Great Britain, e.g., N Devon/E Anglia	3
		Environmental change: a case study, e.g., People-environment interaction in N Norfolk; tourist development e.g., Languedoc – Roussillon; Rhine Delta/Lake Ijssell schemes; Aswan Dam and the Nile Valley; diversion schemes in central USSR.	7, 8

(LEAG A)

Check whether the places and regions mentioned in your syllabus are, like these, only suggestions or if you are required to study them.

VARIATIONS

There are two main exceptions to this account of what you can expect in your GCSE Geography syllabus.

The first is the syllabus which is organized in a quite different way with the main sections being geographical *regions* rather than themes:

B: THE EUROPEAN ECONOMIC COMMUNITY (EEC)

Population and settlement

CONURBATIONS, ILLUSTRATED BY A STUDY OF PARIS

Urban growth and change. Congestion, decay, pollution. Differing land values. Planning needs in an established conurbation. The Paris Regional Plan – its aims, solutions and shortcomings.

Agricultural systems

(a) INTENSIVE AGRICULTURE IN DENMARK

Location. Climate. Varying soil characteristics in a glaciated lowland. Economic reasons for the development of the system. Co-operative farming. Changes in farm size, machinery and labour force. Impact of EEC on farming and markets.

(b) PEASANT FARMING IN SOUTHERN ITALY

Traditional methods and characteristics. Response to environmental constraints – relation to relief, climate and soils. Recent changes.

Energy and natural resources and Secondary industry

(a) NATURAL GAS IN THE NETHERLANDS

Location of the Groningen gas field. Its importance to the Netherlands and EEC energy supply. Pipeline links to Paris, Hamburg and Stuttgart.

(b) RIVER MANAGEMENT IN RHÔNE VALLEY (South of Lyons)

Location of the main management stations. The meaning of 'river regime' and 'drainage basin'. The physical characteristics of the river channel – large basins and narrow defiles. Multi-purpose schemes – water control, improved navigation, energy supply, water for agriculture and industry.

(c) THE RŪHR AND ITS CHANGING INDUSTRIAL DEVELOPMENT

Location. Section of the basin showing nature of the coal measures. Rise and decline in coal production. The changing nature of manufacturing. Effect of migrant labour force. Settlement, planning and the environment. Work of the Rūhr Planning Authority.

Tertiary industry

TOURISM IN SPAIN

Location of major tourist areas. Reasons for growth. The nature of the Mediterranean climate and the factors leading to its reliability. Changes in holiday patterns. The effects on the British holiday resort. The development of links between Northern Europe and Spain. Economic importance of tourism. Effect on local communities – advantages and disadvantages to the local economy and the environment.

Transport and trade

(a) THE RHINE WATERWAY

The meaning of the terms 'navigable river' and 'head of navigation'. Importance of river traffic and location of main river ports. Types of cargo, volume and destination. Significance of this trade within the economy of the EEC.

(b) THE GROWTH OF ROTTERDAM/EUROPORT

Reasons for growth – idea of hinterland. Port facilities and functions. Containerization, oil terminals. Links with hinterland by ship, barge, road, rail and pipeline.

(NEA B)

The centre of attention in this case (one of the three sections of that syllabus) is the region. Topics and locations are specified rather than suggested and so questions will be set which require that specific knowledge.

The second exception is the syllabus which is designed wholly or partly by the school or college. It may well be set out in the same way as the syllabus extracts given earlier. But, within the National Criteria, the school or college has the freedom to choose the ideas, skills and locations it thinks are appropriate.

In short, there are no compulsory themes, topics, skills and areas for GCSE Geography. The National Criteria content guidelines are in broader terms requiring, for example, first-hand study of a small area, consideration of the UK's relationships with other nations and study of social and environmental issues. *Which* small areas, *which* 'other nations' and *which* issues vary from syllabus to syllabus and even between different courses on the same syllabus. To find a clear route through the many topics, places and issues you will encounter during your GCSE course, check your syllabus for the skills, ideas and (perhaps) locations which are at the centre of it.

SECTION 5

Which type of assessment?

COURSEWORK

Work undertaken during the course and assessed by your teachers will probably form the minor part of your assessment, though coursework as a proportion of the total marks ranges from 20% to 60%. The type of work required is laid down in the syllabus as are certain conditions about how the work should be done, such as how much teacher assistance is allowed. How the work should be marked is also set out, in the form of criteria for all teachers who do the marking to apply. While such rules are there for all candidates to observe, decisions on which particular work to do and when and how to do it are for you and your teacher. When writing answers to examination papers, you are confined by the limits of a question and of the space and time you are given in which to answer. In contrast, most coursework allows you to do something which demonstrates what you can do when freed from those limits.

Each GCSE Geography syllabus has a different set of coursework requirements. The commonest is the *fieldwork project*. In some cases two or three such studies are suggested as an alternative to the one project on a single topic. A second type of coursework is a series of *coursework exercises*. One syllabus, for example, requires **'two *separate* and *different* pieces of work covering a range of ideas and questions from either of the option modules' (WJEC/MEG Avery Hill)**. The third type of coursework are *tests* set by the teacher at intervals during your course. Such tests are the type of coursework closest in form to a final examination paper. However, the teacher is able to set and mark test exercises which relate much more closely to the work in your school or college than the questions set by examiners for all candidates at the end of the course.

More information about coursework, and advice on how to tackle it, is given in sections 6 and 7.

EXAMINATION PAPERS

Your GCSE syllabus will state whether there will be one or two examination papers, how long you will have in which to answer them, how many questions will be set and what choice of questions (if any) will be offered. Also indicated in the syllabus and illustrated on the specimen examination papers is the type of question to be set.

You can expect to answer some *structured questions* as all GCSE examinations make some use of that type of question. The question on unemployment at the beginning of the book (pages 1–3) shows what a structured question looks like – a series of linked but distinct parts of varying length and mark weighting, many of the tasks based on information supplied in a variety of forms. In addition to structured questions, a few syllabuses use other types of questions. These include *objective questions*, to answer which you choose from the several possible answers suggested, *short answer questions* and a *problem-solving exercise*. For more about each type of question and how to answer them, turn to sections 8 and 9.

A choice of papers?

Before thinking about questions in more detail, there is one important decision you may have to take. All GCSE Geography syllabuses have a *common course content* which all candidates study, irrespective of their ability in the subject. With many syllabuses, while questions within the exam papers will vary in difficulty, all candidates also tackle the same papers. If such *common papers* are the rule for your syllabus (all NEA and SEG syllabuses; LEAG A, LEAG B; and the Avery Hill syllabus offered jointly by WJEC and MEG), you can ignore what follows and go straight to section 6 on page 32.

It may be, however, that your syllabus provides for a choice of which examination paper to take, the various papers being set on the same topics but at different levels of difficulty – a system of '*differentiated papers*'. With all the MEG syllabuses you have to make a choice of one from two or three such papers. The MEG A syllabus, for example, explains the system in this way:

Candidates should be entered for *one* of the papers described below.

The duration of each paper will be 2¼ hours.

EITHER

Paper One

This paper is targeted at those candidates who are expected to achieve grades D to G.

All questions should be answered. The number of questions set will be unspecified but they will be structured to allow short answers.

OR

Paper Two

This paper is targeted at those candidates who are expected to achieve grades C to E.
All questions should be answered. The number of questions set will be unspecified, but there will be scope for both short answers and free response writing.

OR

Paper Three

This paper is targeted at those candidates who are expected to achieve grades A to D.
Question one and two are compulsory. These test mainly map and graphic skills. Candidates must then choose any *three* from the remaining five questions, which will require mainly free response answers.

(MEG A)

For the Welsh (WJEC) and LEAG D syllabuses there is one common paper for everyone plus a choice between a further two papers, one more difficult than the other. In Northern Ireland, everyone takes two common papers with a third paper being optional for those aiming at the highest grades.

To come out of an exam of this kind with the grade you deserve you will need not only, as with any exam, to do yourself justice on the day, but also to judge in advance which of the various papers will give you the best chance. That judgement will have to be made several months before the date of the examination since you will be entered for only *one* of the alternative papers.

You could see the choice as a straightforward matter of 'how good am I at geography?', but even a glance at the question papers will show it is not as simple as being 'good' or 'weak' at the subject. Look at the two parallel questions on the following pages from different papers on a broadly similar topic. The first is taken from Paper I of the MEG A syllabus referred to above:

Study Figure 9, a model of a city in a developing country.

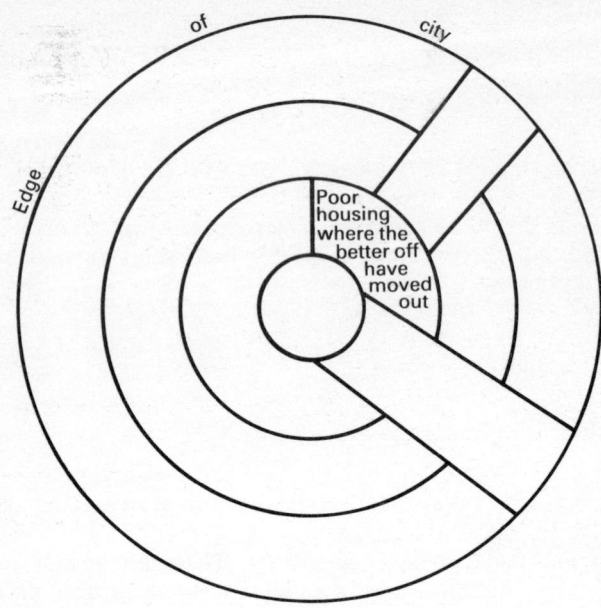

Figure 9

(a) Mark on Figure 9
  (i) CBD where you would find the shopping centre and office
       blocks. *(1)*
  (ii) F where modern factories have been built along main roads.*(1)*
  (iii) S where squatters have built shanty towns. *(1)*
  (iv) X where you would find expensive, modern high rise flats. *(1)*
  (v) P where you would find poor quality but permanent houses,
       most of which have piped water supply; some are connected to
       the sewer. *(1)*

(b) Name a city with shanty towns, which you have studied. *(1)*

(c) What attracted people to move to this shanty town? Suggest *two*
      attractions. *(2)*

(d) What was bad about the area they used to live in? Suggest *two*
      features. *(2)*

(e) Describe living conditions in the shanty town you have studied. *(2)*

(f) Why are people prepared to live in such conditions? *(1)*

(g) Imagine you are the Mayor of the city with shanty towns which you have studied. You have a little money to spend on the shanty town.

    (i)   What improvement would you spend it on?

    (ii) Why? *(1)*

                            (MEG A)

The second question is taken from the most difficult of the three MEG A specimen question papers, Paper 3 (Fig. 5, referred to in the question, is on page 30):

Study Fig. 5 showing population growth in the world's major cities.

(a) Explain the differences in growth rates of cities shown in the developed world and the developing world. *(6)*

(b) Shanty towns are a feature of many of the fastest growing cities. Describe *one* example you have studied. *(6)*

(c) Explain how the quality of life in a shanty town differs from that in:
    (i)   an inner city area in Britain and
    (ii) in a New Town in Britain. *(8)*

                            (MEG A)

A comparison of the two questions illustrates some of the differences you can expect to find between papers at different levels of difficulty.

Paper 1 has a larger number of questions, each requiring a relatively brief answer for one or two marks. Paper 3 asks fewer questions but expects short-essay-type answers to each. The diagram supplied for Paper 1 should be easier to interpret than the larger and more complex set of data on the world map in Paper 3. Many of the marks for the Paper 1 question can be obtained by remembering what you have learned about shanty towns, whether it be the typical location of certain features *(a)* or the name of an example *(b)*. For Paper 3, however, not only do you have to write at greater length, you are also expected to use what you know to *explain* the growth rates on the map *(a)* and to *compare* the quality of life in a shanty town and in urban areas in Britain *(c)*.

The same points of contrast between Paper 1 and Paper 3 on the MEG A syllabus can be seen if you compare the question on Mount St Helens (page 71) with that on UK water strategy (page 73). From these examples the choice *you* would make may seem clear enough. But remember that there is also a Paper 2, intermediate in difficulty between

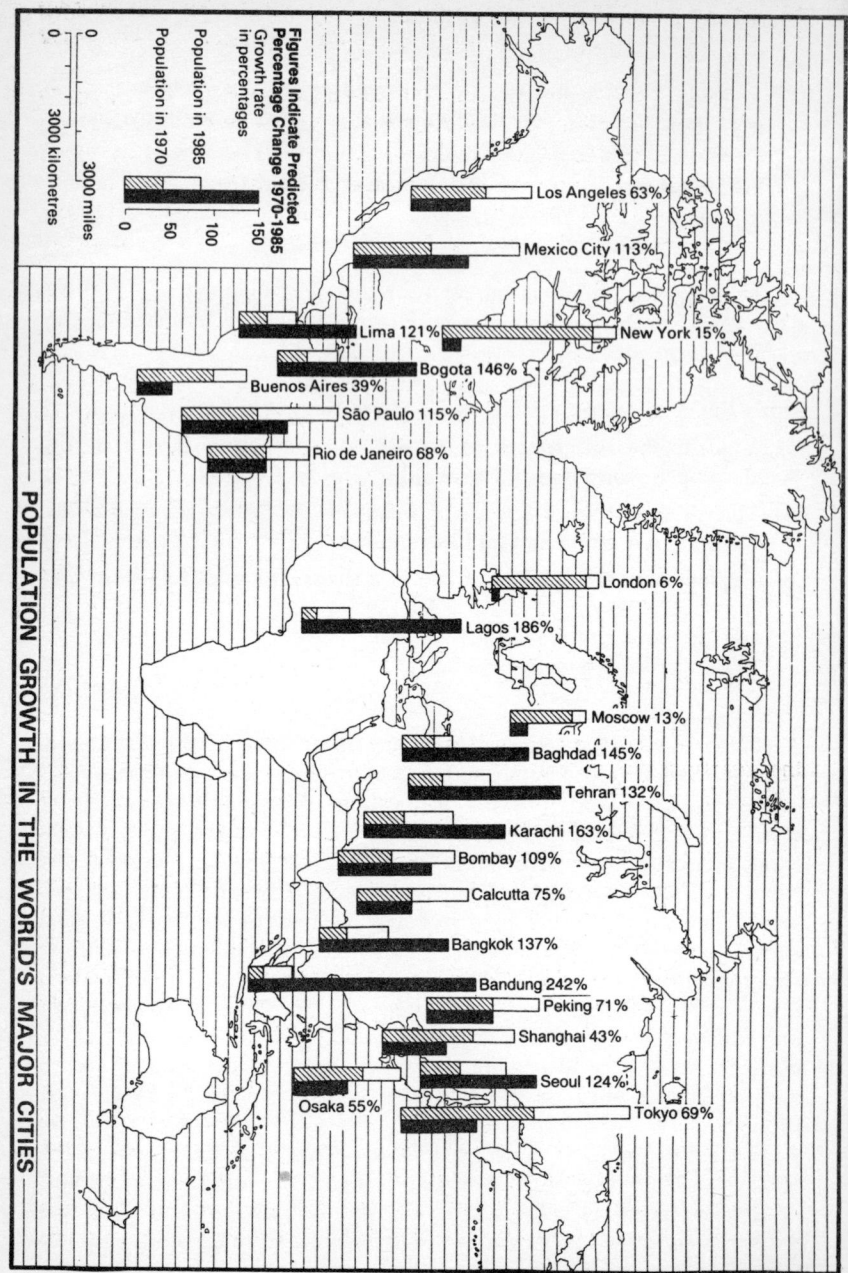

Figures Indicate Predicted
Percentage Change 1970-1985
Growth rate
in percentages

Population in 1985
Population in 1970

0   50   100   150

0   3000 miles
0   3000 kilometres

Los Angeles 63%
Mexico City 113%
Lima 121%
New York 15%
Bogota 146%
Buenos Aires 39%
São Paulo 115%
Rio de Janeiro 68%
London 6%
Lagos 186%
Moscow 13%
Baghdad 145%
Tehran 132%
Karachi 163%
Bombay 109%
Calcutta 75%
Bangkok 137%
Bandung 242%
Peking 71%
Shanghai 43%
Seoul 124%
Osaka 55%
Tokyo 69%

POPULATION GROWTH IN THE WORLD'S MAJOR CITIES

Fig.5

the papers from which these examples are taken. The differences between alternative papers are not always as marked as in these examples.

As well as the evidence from the specimen question papers as to what type of question you could expect and what sort of answer you need to produce, the MEG A syllabus gives another clue as to which paper you should choose. It tells you that, with 50% of the marks on Paper 1 you could expect a grade F, 50% on Paper 2 would bring a grade E, while a similar level of success on Paper 3 would probably reward you with a grade C.

How then do you make the right decision? If you select a topic from your syllabus and look at the questions set on that topic in each of the alternative papers, you will find some typical differences. The paper on which the lower grades are awarded is likely, among other things, to have questions which are more simply worded, demand less understanding of the topic and require shorter, more direct, answers. Conversely, the paper on which a good performance is required for a higher grade to be awarded is likely to ask more complex and demanding questions and expect you to write at greater length showing more understanding. There may well be a variation of difficulty *within* the questions on each paper but, taken as a whole, the equivalent questions on the two papers will differ in the ways suggested here.

Which of the alternative papers to opt for has been dealt with at some length because it could have a major effect on the grade you are awarded. You may be by nature optimistic and/or ambitious and be inclined to aim high. Before you do so, bear in mind that, if many of the questions prove too difficult for you, the examiners could be left with so little evidence on which to base an award that, far from the A you had hoped for, you might be left with no grade at all. Conversely, you may tend to underestimate yourself or prefer to play safe and settle for the easiest paper. If you then do outstandingly well, you would have to hope that the examiners would be willing to award a grade higher than the maximum normally awarded to candidates taking that paper. In practice, examiners of the MEG A syllabus might, exceptionally, award a C to a candidate taking Paper 1 but you could not expect an A if you opted for that paper, however good your answers.

Take stock before deciding which paper(s) you should opt for. Be realistic about the type of question which, given some revision, you could expect to be able to answer. Weigh up your performance in any 'mock' GCSE examination you may take. Discuss it with your teacher(s). Together you should be able to agree on the question papers which offer you the best prospect of demonstrating what you know and can do.

SECTION 6
Understanding coursework

The written work you do during your course will usually be marked by your teacher. For the most part the marks given for class exercises, tests and end-of-term exams are for use only within your school or college as indicators of how well you are doing. But every GCSE course includes a few pieces of work (your teacher will tell you which) that count towards your GCSE grade. It is this coursework – it could be several exercises at intervals through your course or perhaps one major project – which is explained here.

The three questions to ask yourself are:

What exactly am I required to do?
How will I be judged on what I do?
How can I write good coursework?

This section helps you answer the first question; the next section deals with the other two.

Unlike the final exam papers, there is no one series of exercises which every candidate has to attempt. Within the rules set out in the syllabus, the choice of coursework is left to your teachers and, in some cases, to you.

Yet too often good coursework is given low marks because, though good in its own way, it does not fit within these rules. There should be no mystery about what is required: it is usually spelled out in great detail in the syllabus. However, it is all too easy, for example when you are busy collecting information for a project, to forget the rules which govern the choice of project, how it should be done and the way marks will be awarded. Inform yourself about such things first and then you will be able to concentrate on doing the work as well as you can.

The table opposite shows which of the three main types of coursework are required for each GCSE syllabus. The three types are:

1 *Fieldwork projects* (variously called individual studies, geographical enquiries, geographical investigations, course studies).

2 *Coursework exercises*

3 *Course tests*

	MEG				NEA				LEAG				SEG		WJEC		NI SEC
	A	B	C	D	A	B	C	D	A	B	C	D	A	B		†Avery Hill	
Fieldwork projects (page 34)	1 or 2 or 3		2	1 or 2	1 or more		2	1 or 2	2	3	1 or 2	1	1	1	1	1	1
Coursework exercises (page 41)			2	3				3						2		2	
Course tests (page 44)						4*											
Coursework as a % of total marks	25	28	50	50	25	60* or 25	30	50	30	40	20	25	25	40	20	40	20

Coursework: number of exercises required of each type.

* Course tests optional in place of one exam paper
† Avery Hill syllabus offered jointly by WJEC and MEG

33

Use the table to find your own route through this section, reading only those parts which are relevant to your syllabus.

FIELDWORK PROJECTS

You have probably completed several projects in different subjects during your school career. But beware of assuming that what is expected of you for a GCSE Geography project (or projects) is more of the same as was expected when you studied dinosaurs in junior school or The Romans in first-year History. GCSE Geography projects are much more than an open invitation to find out about a chosen topic. For the examiner, they are a chance to test you on abilities which cannot be tested within the limits of the examination room. To be fair to all candidates, the examiner must state clearly the type of work that is required.

Doing one or more projects gives you the opportunity to develop your own interests and to apply the skills you learn in geography. For those reasons the project is for many people one of the most enjoyable and satisfying parts of the course. At first sight, the scope for what you could do seems limitless. A list of possible questions for enquiry hints at the range of possibilities:

Examples of student-planned enquiry titles

These have been stated as questions, but could equally have been framed in terms of a hypothesis, e.g. *18* **House prices increase with distance from the city centre.**

1 **How do soil characteristics change down slope x and why?**

2 **How does the nature of the vegetation change from x to y and why?**

3 **How does the nature and size of the bed load vary in stream x from upstream to downstream and why?**

4 **How does the surface velocity vary within stream x and why?**

5 **How do river channel characteristics vary from upstream to downstream/between a meander and straight stretch in stream x and why?**

6 **To what extent is stream x polluted and why?**

7 **What factors influence the variety of cliff profile from x to y?**

8 **How do beach/river/morainic/scree deposits vary in size and shape and why?**

9 **How do slope profiles/valley cross-sections vary from x to y and why?**

10 How do the effects of an anticyclone and a depression influence the weather recordings at station x and why?

11 What differences occur in the micro-climate of a woodland area and an area of open land/within a school building area and why?

12 What effects do variations in rainfall have on the discharge at a river station(s) at x and why?

13 How and why do levels of litter/vandalism vary in town/area/village x?

14 How do a coniferous and deciduous wood compare in their impact on the environment and why?

15 To what extent does the knowledge and perception of a place vary with age, social class and distance and why?

16 To what extent does land use vary in the CBD of city/town x and why?

17 To what extent does the quality of environment vary in town x and why? Can environmental quality be related to socio-economic characteristics of each area?

18 How do house prices vary in area x and why?

19 What factors influence the distribution of crimes and criminals in town x?

20 How do shopping patterns vary between a hypermarket and/or CBD centre and/or neighbourhood shopping centre and why?

21 What factors influence the size and shape of the hinterland of town x?

22 To which major centre do the people of village x look towards and why?

23 What factors influence the pattern of land use on farm/area x?

24 In what ways has village x changed since its beginning and why?

25 What factors influence the holiday habits/leisure activities of the population of area x?

26 To what extent can the range of services offered by a settlement be related to its population size?

27 Are the leisure facilities of area/town x adequate for the needs of the people?

28 In what ways has the industrial structure and pattern changed in town x and why?

29 What factors have influenced the location of industry/industries x in area y?

30 **What impact has tourism had on town/area x and has this impact been good or bad?**

31 **Where are the main traffic congestion points in area x and why?**

32 **To what extent is village/town area x a commuter settlement?**

(NEA D)

But before becoming intrigued or perhaps worried by the breadth of choice, you should recognize that the options are not as open as they might seem. GCSE syllabuses vary not only in the number of projects required (see the table on page 33) but also in the type of project which will be acceptable. Your teacher will guide you as to what your syllabus asks for. To work it out for yourself, use the following questions as a checklist to give a clear picture of what you will have to do.

A free choice of topic?

Usually not. A project is typically seen as an opportunity to extend your study of topics on the syllabus. So you may find clear limits on your choice. For example:

Each Enquiry should be drawn from a different theme in the syllabus, i.e., Urban Geography, Economic Geography and People and the Environment. Each should also be based on or developed from the key ideas in the syllabus content.

(NEA C)

Other syllabuses are less prescriptive:

Suitable subjects for Geographical Enquiry should consider the issues to which the key ideas relate but they may go beyond the content stated.

(MEG A)

Which sources of information?

For many project topics, relevant information could be collected either from your own first-hand observations ('primary data'), or from second-ary sources such as books, statistics or maps. While collecting your own information in the field can be very enjoyable and give you a personal insight into your topic, there are some situations where it makes sense to rely on published material such as stream flow data, population census returns or weather statistics. Given a free hand, and depending on your chosen topic, you might find both primary and secondary sources useful.

But one standard feature of all GCSE Geography syllabuses is that field study must form a part of the assessed coursework. So, for example:

Secondary source material may be used to *supplement* the information obtained by first-hand enquiry but a submission based entirely on secondary source material is not acceptable.

(MEG A)

Check carefully whether you must base your project (or projects) wholly or partially on field study.

Which methods of study?

To focus your attention within a topic, it is helpful to think of a question or questions which you hope to answer through studying it. Indeed the list of suggestions given earlier was put in that way. **How do soil characteristics change down slope X and why?** gives you something to work on which is clearer than a topic such as 'a study of soil characteristics'. Many syllabuses do more than just suggest that you should think this way: they require it. For example:

[**The Geographical Enquiry**] must involve the consideration of an argument or a problem or an assertion to be tested.

(SEG A)

So, for instance, instead of putting forward 'a study of recreation facilities' in your town you are encouraged to deal with the same topic in a particular way, such as arguing that a certain area of the town has poor facilities or perhaps proposing where a new leisure centre might be sited.

Beyond starting with a question, problem or hypothesis, you may also be required to carry out a project in a particular way. Several syllabuses include something similar to the 'general approach to geographical enquiry' shown on page 38:

(NEA A)

It would be unwise to ignore such guidance since, where it is a requirement, marks are allocated as to how effectively you have used the mode of enquiry. The same requirement may also limit your choice of project. Some potentially interesting project exercises, such as writing an 'alternative' town guide or devising a development plan for an area of derelict land, would not be compatible with using this specified method of 'scientific' enquiry.

Which techniques?

The techniques you use will obviously vary with the topic. You could find yourself wearing waders and using a stream flow meter or armed with a questionnaire and clip board interviewing shoppers. Your teacher will suggest suitable ways of studying your chosen topic.

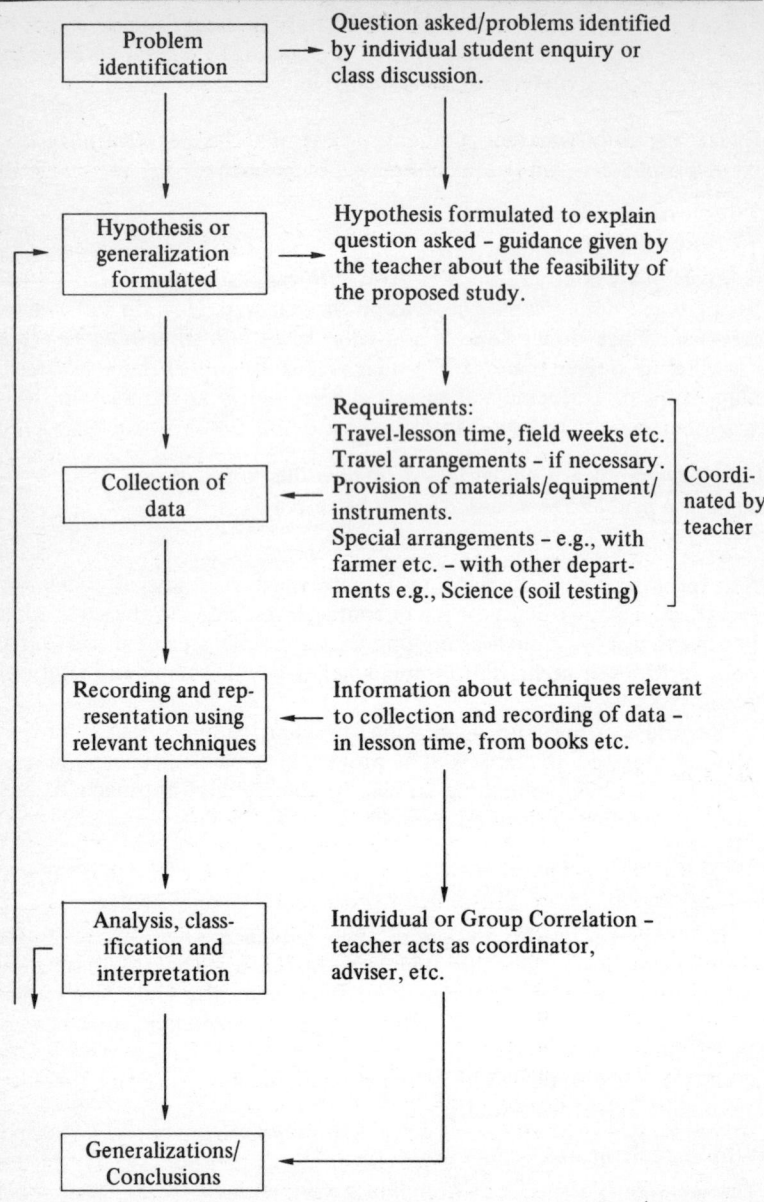

```
┌──────────────────┐     Question asked/problems identified
│     Problem      │ ──→ by individual student enquiry or
│  identification  │     class discussion.
└──────────────────┘
          │
          ↓
┌──────────────────┐     Hypothesis formulated to explain
│  Hypothesis or   │ ──→ question asked – guidance given by
│  generalization  │     the teacher about the feasibility of
│   formulated     │     the proposed study.
└──────────────────┘
          │
          ↓
                         Requirements:
                         Travel-lesson time, field weeks etc.     ┐
                         Travel arrangements – if necessary.      │ Coordi-
┌──────────────────┐     Provision of materials/equipment/        │ nated by
│  Collection of   │ ←── instruments.                             │ teacher
│      data        │     Special arrangements – e.g., with        │
└──────────────────┘     farmer etc. – with other depart-         │
                         ments e.g., Science (soil testing)       ┘
          │
          ↓
┌──────────────────┐     Information about techniques relevant
│ Recording and rep-│ ←── to collection and recording of data –
│ resentation using │     in lesson time, from books etc.
│ relevant techniques│
└──────────────────┘
          │
          ↓
┌──────────────────┐     Individual or Group Correlation –
│ Analysis, class- │     teacher acts as coordinator,
│ ification and    │     adviser, etc.
│ interpretation   │
└──────────────────┘
          │
          ↓
┌──────────────────┐
│ Generalizations/ │ ←──
│  Conclusions     │
└──────────────────┘
```

(NEA A)

Though the syllabus may not specify techniques you must use, it will make clear the different *types* of techniques to be used in completing your project. As one syllabus puts it, you should be able to:

in relation to skills, select and use a variety of techniques appropriate to a geographical enquiry, including investigation in the field and, in particular to

(i) use basic techniques for obtaining, observing, recording, representing, classifying and the critical analysis and interpretation of evidence;

(ii) use and analyse a range of source materials, including audio-visual materials, maps at a variety of scales, graphs and simple statistical data;

(iii) demonstrate an ability to problem solve by selecting, using and communicating information, ideas and conclusions in a variety of forms;

(Avery Hill)

Put more briefly, you must choose the right techniques for collecting, analysing and presenting information relating to your project topic. Finding the evidence you need or presenting the results in a neat series of graphs is not enough on its own. Make sure you know not just how to collect the information but also how best to process and to present it.

All my own work?

Group projects are normally allowed, sometimes encouraged. The one condition is that the attribution of who did what is sufficiently clear for the teacher who marks the project to be able to tell what each member of the group contributed. If this option is open to you, consider it seriously. If you and a friend or friends share an interest, collecting material together can be more enjoyable than working on your own. Sometimes, individual project work can be developed from work which a whole class undertakes together.

Then there is the question of advice and help from your teacher. Your teacher is supposed to help you decide on a project and later to mark it; he/she is not supposed to write it for you. To reduce unfairness to candidates from different schools, there are limits to the assistance teachers are allowed to give. Beyond a certain point in the preparation, usually when the information has been collected, you should work more or less on your own.

Two GCSE syllabuses distinguish student-planned enquiries (SPEs) from teacher-planned enquiries (TPEs), with candidates doing one SPE or two TPEs. The division of labour between yourself and the teacher is shown in this way:

The enquiry(ies) should enable candidates to carry out investigations using the following route to enquiry.

		SPE	TPE
	Identification of issues, question, problem.	Individual student discusses idea with teacher.	Class discuss work and the planning of the enquiry.
	↓		
	Formulation of design of work.	Individual student discusses feasibility, scale, resources, etc.	Teacher assembles resources, organizes group.
	↓		
Use of primary sources →	Collection of data, first-hand recording, sketching, mapping, etc.	Individual carries out fieldwork.	Class carries out group exercise.
	↓		
	Selection and collation of data.	Individual student selects, collates data with guidance from teacher.	Teacher collates data for class use, students select data to answer question.
	↓		
	Representation and reading of results.	Students individually record results and represent findings by maps, graphs, etc. Teacher guidance on techniques.	
Use of secondary sources →	Analysis and interpretation.	Students individually analyse and interpret their findings in answer to question.	
	↓		
	Making effective conclusions, suggesting solutions, evaluation.	Students individually conclude the results of their findings and evaluate their findings.	

(NEA D)

How about help from family or friends? Clearly ideas and suggestions can come from anywhere, but you will have to sign a statement that the project was completed 'without any external assistance' except the teacher assistance which the syllabus permits.

How to present the project?

You are limited as to length of write-up, usually between 1500 and 2500 words. Try to stick fairly closely to the required length. It will be obvious that an under-length project could be penalized for its inadequacy. It may not be as obvious to you that an over-length report, however impressive a production it may be, could be looked upon unfavourably by the marker. Your teacher, faced with a whole set of projects, will not thank you for saying what you have to say at undue length.

Some requirements have to do with the examiners' convenience – 'the work of each candidate should be presented neatly in a paper or manilla folder, not a ring file' (WJEC). Such regulations should not be ignored. You may even find the form of your written report of the project stipulated, as for example:

Each presented study should be in the form of an *A4 folder*, labelled with Subject Name and Number, Centre name and number, candidate's name and number, study title, and containing the following:

(i) A table of contents
(ii) A brief introduction to state the aim of the enquiry
(iii) Methods/data collected
(iv) Data and findings
(v) Analysis of data
(vi) Interpretation and conclusion
(vii) Bibliography (if relevant)
(viii) Appendix

(SEG B)

COURSEWORK EXERCISES

The basic difference between coursework exercises and a project is that all students in the same school or college undertake the same coursework. The tasks are built in at appropriate points within the normal course programme. You will do the work as if it were ordinary exercise work but, unlike most of the work done during the course, the marks given by your teacher will count towards the final grade. Some syllabuses stipulate the length of time that the work should take – typically about two weeks – but others allow the teacher to decide whether to set a series of short tasks or to assess your work across a whole section of the syllabus.

There is considerable freedom for a teacher to decide what exercises are given. In some cases the only clear rule is that the work should be different from the kind you will be assessed on in the final examination. For example:

The assessments may take different forms:

(a) Continuous assessment throughout a module.

(b) Assessment at the completion of a module.

(c) Individual work by candidates as part of the study of a module.

The coursework assessment should be regarded as an opportunity for employing methods of study and presentation which do not lend themselves to the time restriction or style of the final examination. For example, the candidate may be required to tabulate and analyse a large quantity of statistical data or to assimilate extensive stimulus material including videos or interviews or the construction of detailed maps and diagrams or participation in geographical games or simulations.

(SEG B)

Other syllabuses limit the topics to be included and suggest examples. One syllabus requires a coursework unit in each of three categories:

(a) The study of the physical environment

(b) Planning problems within natural and/or built environments

(c) Area studies

(NEA D)

For the first of these, the syllabus guidelines are:

(a) The study of the physical environment

This unit must allow the student to examine the operation and interaction of physical processes, for example in the development of landforms, weather and climate, soils or vegetation. At the same time it is envisaged that part of the assessment could include the impact of physical processes on people or their attempts to manage the natural environment.

The following are examples of relevant studies:

(i) The causes and effects of a selected natural hazard e.g., volcanoes, hurricanes, landslides.

(ii) An analysis of the distribution, nature and possible reasons for formation of selected landforms for example along the course of a stream/river, at the coast or in a glaciated area.

(iii) Climatic variations within a chosen area such as the school grounds.

(iv) Variations in the nature and usefulness of soils in an area, for example a catena down a slope.

(v) Nature and formation of vegetation within an area, with the possible analysis of people's impact on the ecosystem.

(vi) An investigation of the hydrology of a small river catchment.

(NEA D)

In another case (MEG C), the syllabus does not specify the topics of coursework exercises but rather the type of task you will be given – a decision-making exercise. For that you will be given information about a geographical issue and be expected to use your geographical skills and judgement to reach and justify a decision. An example will illustrate what is required.

After having had about a week to study the four information sheets relating to choosing a site for a 'Hi Tech' industry, you would be given a maximum of an hour to answer the questions set out on a task sheet:

TASK SHEET

Read the following instructions and guidelines thoroughly.

Introduction

In this exercise you are to imagine that you are a Chief Executive of a successful European 'Hi Tech' electronics company.

The company wants to expand in the UK but has yet to decide *where*. The company wants you to find out about a possible location site in the UK because

(a) it will be in a better position to serve its existing UK customers.

(b) it wants to try to attract more UK orders.

Your task is threefold

1 In this task use Information Sheet One.

 You have to go back to your company headquarters and describe to the shareholders the present distribution of 'Hi Tech' electronic firms in the UK and suggest reasons for this distribution.

2 In this task use Information Sheets Two and Three.

 You have narrowed down your choice of a site to *five* possible locations.

 (a) Bradford
 (b) Cumbria
 (c) Livingston
 (d) Mid Glamorgan
 (e) North Kent

 You have now to explain to the shareholders *briefly* what you consider to be the main advantages and disadvantages of each location.

3 Using all the information you have been given (including Information Sheet Four), which of the five possible locations do you consider to be the best site for your future UK factory?

(MEG C)

Interestingly, the purpose of this is very similar to that of the problem-solving exam paper used in another syllabus (see section 8) – to give you a chance to *apply* what you know and can do. The difference between the two is a good illustration of the two major differences between exam questions and coursework. The decision-making exercises referred to here are chosen, and later marked, by your own teacher and they allow more time and scope for you to show your abilities than would a question to be answered within an hour in the examination room.

COURSE TESTS

For almost everyone, tests set during the GCSE course will serve the usual purpose of indicating how well you are doing. The results will not affect your final grade. However, for one syllabus (NEA B) four tests may be set and marked by the teacher with the marks counting towards the final grade. The questions will be different from those set in the final examination paper. For example:

1 A topic research by an individual candidate or by a group of candidates and tested in class under examination conditions.

2 A question in which the candidates are allowed to use any material provided by the teacher e.g., textbooks, resource material, photographs, maps, etc.

3 A test of understanding of material contained in video, slide and film, etc.

4 A test of geographical skills e.g., the construction of charts, graphs, etc.

5 An oral test e.g., based on geographical material to show the candidate's understanding.

(NEA B)

SECTION 7
Writing coursework

When writing coursework you have two significant advantages compared with going into the examination room. First, with examination papers, you can only guess at what will be given credit on the questions you are answering. However, with coursework, what counts as success is published in advance in the form of a marking scheme. Second, with examination papers, it may be difficult to read the intentions of an unknown examiner who sets the questions and marks your answers. With coursework, the work is set and marked by your own teacher. Whatever type of coursework you are doing, make sure you take full account of both these factors to ensure that you show your abilities in ways that are recognized as credit-worthy.

THE PROJECT

How it is marked

The notion of thousands of candidates writing projects on a huge variety of topics to be marked by hundreds of teachers presents obvious problems of fairness to the examining groups. One result is the detailed set of rules, which were explained in the previous section, about what is acceptable. The other is their clear statement as to how the marks for projects will be allocated.

For your particular syllabus you may find the guidelines on marking are as general as this:

Criteria for assessment of geographical enquiries   *Marks*

(a) **Collection of primary data, and where appropriate, supporting secondary data relevant to a topic.**   *12*

(b) **Presentation of data using a variety of geographically appropriate forms.**   *12*

(c) **Analysis and interpretation of data by application of geographical concepts and principles, including identification of values and their role in decision making.**   *14*

(d) Conclusions drawn from the findings of the enquiries, including, where appropriate, proposals, justifications and evaluations for solutions to geographical problems.　　　*12*

　　　50

(MEG A)

Like all the marking schemes, this underlines the importance of giving due attention to each stage of preparation and the writing-up of your project.

You may find there are more clues as to what standard is expected of you, such as these:

CRITERIA FOR MARKING TEACHER- AND STUDENT-PLANNED ENQUIRIES

(a) Posing the problem

　　5 A very clear definition as to the purpose of the enquiry with an appropriate selection of route enquiry and good understanding of the wider implications of the issue.

　4–3 A sound attempt to define the purpose of the enquiry and to summarize the key issues.

　　2 An adequate description as to the purpose of the enquiry, as described by the teacher.

　　1 A basic understanding of what the enquiry is trying to achieve.

(b) Collection of the evidence/data

　10–9 Evidence of very thorough observations, accurate measurements and recording of a sample size appropriate to the enquiry, using a wide range of practical skills and personal initiative.

　8–6 Evidence of thorough observations and accurate measurement and recording, showing some initiative.

　5–3 Evidence of sound observation, measurement and recording, by following instructions.

　2–1 Evidence of basic recording and collection of data, after close guidance.

(c) Organization and representation of the evidence/data

　10–9 A very well/logically organized study containing a range of well-presented maps, diagrams, graphs, etc. completely appropriate to the task and showing considerable initiative in choice and design techniques.

8–6 A well organized study containing a range of neatly presented maps, graphs, diagrams, etc.

5–3 A basic, partly organized account, in which all the results are graphed or mapped with some reliance on teacher guidance.

2–1 A study in which the results are graphed or mapped by simple means, possibly using provided frameworks.

(d) Analysis/interpretation of the results of the enquiry

10–9 A very good analysis of the results, including, where appropriate, simple statistical techniques and an interpretation of the results, showing considerable initiative and insight.

8–6 A thorough analysis of the results, with some interpretation of their significance.

5–3 Clear description of what the results show.

2–1 A very limited description of the findings, possibly using a closely structured format provided by the teacher.

(e) Conclusion and evaluation of the enquiry

5 Clear, concise statement of the outcomes of the enquiry, with a full recognition of the limitations of the enquiry, and/ or some evaluation of its effectiveness and possible application and extension.

4–3 A sound assessment of the outcomes of the enquiry and some evaluation of its success.

2 A clear comment on what the enquiry showed and how useful it was.

1 A brief comment on what the enquiry showed, after staff guidance.

(NEA D)

To take just one instance, if all you attempt at the end of your project is a short statement of the outcome, your teacher cannot give you more than 2 of the 5 marks available in the last category. You must at least *attempt* to do what the highest mark level in each category refers to. Your teacher may decide you have not fully succeeded, but for you not even to try to achieve this level is as foolish as not answering the required number of questions on the exam paper.

Another syllabus expresses the target you should aim for in the project in a different way, by defining the lowest level of achievement, for which up to a third of the 40 marks will be awarded:

Expected levels of performance	Candidates demonstrating attainment at this level could be awarded ...
1 In relation to structured enquiries, given specific guidance by the teacher at all stages, the candidate has demonstrated the ability to: *(i)* collect and record data from primary sources on provided recording material by following precise instructions and using familiar procedures; *(ii)* select relevant information under given headings from a limited number of secondary sources; *(iii)* present the collected data in appropriate forms given step-by-step instructions; *(iv)* present a commentary expressed basically in descriptive terms; *(v)* comment on the application and usefulness of the findings.	*up to one-third of marks*

(MEG D)

In contrast, the level you have to reach to be eligible for more than two-thirds of the marks is explained in this way:

3 Given general guidance on structuring enquiries and occasional tutoring when sought, the candidate has demonstrated the ability to:
 	
  (i) show initiative and imagination in selecting a topic suitable for enquiry;
 	
  (ii) make decisions about geographically appropriate sources of information, strategies for data collection, forms of analysis and cartographic and symbolic means of presenting findings;
 	
  (iii) develop and use a range of practical skills, techniques or equipment to collect and process data from primary and secondary sources;
 	
  (iv) analyse the collected data and draw conclusions and possible implications;
 	
  (v) reflect and comment on the effectiveness of the enquiry, its findings, and possible application and extension;
 	
  (vi) communicate the outcomes of the enquiry appropriately and clearly.
	up to full marks

(MEG D)

The above marking scheme extracts are from three GCSE syllabuses. They have some features in common but are sufficiently different for it to be worth checking your own syllabus's scheme so that you are clear about how marks will be given before you embark on the project.

Some pointers as to how to tackle your project should have emerged from this and the preceding section. A checklist of things to keep in mind at each stage of the work will help to give you a successful result.

Stage 1 : Planning it

It may be helpful to review a list of questions such as that given on page 34 or rather more filled-out ideas such as these:

Examples of possible subjects for geographical enquiry

(a) Location of study

A small old market town with large cattle market and heavily used by tourists in summer both visiting and passing through. Trunk road also passes through centre. A proposed by-pass has been turned down by the Department of the Environment.

Possible subject for study

Candidates could investigate aspects of the conflict between users of roads in the congested town centre investigating, for example, the extent of the problem and the measures being taken to solve the problem, evaluating their success and proposing further methods of tackling the problem.

(b) Location of study

An area of industrial dereliction.

Possible subject for study

Candidates could investigate how the changing nature of industry has affected the quality of the environment. Questions to be considered could include:

What is the extent of a problem?
Have the changes had a negative or positive effect?
How can the negative effects be countered?
How can the positive effects be extended?

An alternative study in an area of this nature might consider the social consequences of change.

A third possibility might be to produce reasoned plans for the development of this land as a country park.

(c) Location of study

A stretch of river bank.

Possible subject for study

Candidates could investigate process in relation to how river channels develop differently. This could include simple levelling and measuring stream velocity across a meander and a section of straight channel.

(d) Location of study

Medium or large urban area.

Possible subject for study

Candidates could test the idea that a shopping hierarchy exists within an urban area. This could include the collection of data on such topics as shop location, shop type, information about shopping habits and perception of services.

(e) Location of study

A National Park.

Possible subject for study

Candidates could investigate aspects of land use conflict in a National Park examining, for example:

The pressures on particular beauty spots resulting from excessive popularity.

The difference of interest between conservationists and water authorities.

The impact of quarrying.

(f) Location of study

A stretch of varied coastline.

Possible subject for study

Candidates could contrast a stretch of coastline where erosion is dominant with one where deposition is taking place. Subjects for study could include:

The measurement of longshore drift and a study of man's attempt to ameliorate its effects.

The explanation of the varied types of distribution of beach material.

The effects of tourism on the coastline.

An investigation of the physical processes which have affected the shape and scale of the cliffs.

(g) Location of study

A large urban area.

Possible subject for study

Candidates could investigate possible locations for a new hypermarket using census data extracted from a computer database. This would be used to establish which wards might have suitable conditions. Possible locations would then be identified and evaluated by first-hand investigation.

(h) Location of study

A stream above and below either a factory outfall or a sewage works or a small to medium-size settlement.

Possible subject for study

The changes in the balance of animal populations with special reference to an indicator species.

An evaluation of water quality.

(MEG A)

However, beware of latching on to a topic which might suit someone else but is not really for you. Ask yourself these questions:

Does it interest me?

If you do have the choice, rather than following what starts as a class exercise, don't make the mistake of choosing the first thing you think of or the last thing you studied in the course. You are going to have to work at the topic over a period and see it through to completion. There are many possible starting points for your thoughts on this, apart from what you have dealt with directly in the geography course. For example:

(i) a local issue such as the location of a site for travellers;

(ii) a rich source of geographical data. Newspaper weather reports and forecasts could lead you to ask what weather patterns can be detected or how accurate are the forecasts;

(iii) a personal interest or hobby which could be linked to the syllabus you are studying, such as the provision for cyclists in an urban traffic plan.

Need it be original?

An unusual topic can cause the marker to sit up and take notice. But his or her initial interest will soon turn to criticism and low marks if your topic is so ambitious that you are unable to find the right information or reach sensible conclusions. Studying the place of origin of tourists during your family holiday in Majorca might sound like a bright idea, but a touch of sunstroke or diarrhoea at the crucial time might ruin your project as well as your holiday.

Will it be acceptable to the examiner?

You should be able to make your own judgement on this if you have read section 5 in conjunction with the relevant part of your syllabus. There should not be a problem, however, because your intentions, once clear, will have to be approved before you begin the work. In some cases it is for your teacher to decide whether your proposal is acceptable. Other syllabuses require a list of suggested projects to be sent to the examining group for approval.

Is it clear how I will tackle the topic/question/problem?
It is easy at the planning stage to become too preoccupied with the subject matter for your project and overlook the practicalities of actually studying your chosen topic. Make sure before you settle on a subject that you have a good idea of where you will get your information and which methods you will use to analyse that information.

Will I be able to finish it?
Ask yourself if, given the topic you have in mind and the type of project you have to submit, you will be able to achieve definite results in the time available. Once you have settled on a topic and the way of tackling it, think about halving its scope and then halving it again. You may then be nearer to a manageable enterprise. It is an almost universal tendency when anticipating writing a project to overestimate what can be achieved in the time available. Only when you are into the work does it become obvious how much information there is and how long it takes to gather and process that information. As one syllabus puts it:

Studies should be clearly defined and straightforward. Carefully stated hypotheses/problems/assertions are needed.

(AVERY HILL)

For example, you may be interested in the characteristics of rivers but how much information can you expect to be able to handle? The four suggestions given in the list of possible topics on page 34 (3 to 6) is an indication of how narrow the focus may need to be for a GCSE project. Concentrating on one element of the stream – bed load, surface velocity or pollution – gives you something to fix your attention on and a manageable amount of data to process.

Stage 2: Carrying out the work
How you tackle the project will depend a great deal on how individual it is, your choice of topic and the requirements of your syllabus. However, keep these general points in mind:

Use appropriate techniques in the right way

It is the relevance and quality of the information you collect which matters, not the quantity. Although you will be assessed partly on whether you have used techniques correctly, what is just as important is that they are the right techniques for your chosen purpose. Try to collect information which will provide positive or negative answers or lead to clear-cut conclusions (though the conclusions should not be so obvious that they could have been seen at the outset).

For explanations of possible techniques (and suggestions for project topics) you may find these books helpful:

Jennifer Frew, *Geography Fieldwork*, Macmillan Education;

Brian Greasley, *Project Fieldwork*, University Tutorial Press.

Plan when you will do what

You will be given a deadline date by which the finished project must be submitted. You may work best when that date is almost upon you or you may prefer to arrange things so that the writing-up is completed in good time. Whether writing up the results should be left to the last minute is for you to decide. What is certain is that you will not be able to *collect* and *process* the information, especially if it is to be collected by field study, the night before the project is due to be in. A timetable for when you do what is essential.

Stage 3: Writing it up

Don't just sit down and write the 1,500 or 2,000 words required without giving some thought to the presentation of your report. However good the work you have done, remember it is the written report of that work on which you will be assessed. Before writing up your project, think about:

A clear line of argument

All studies should be introduced in terms of clear objectives, data to be analysed and method of approach. It should not be necessary . . . to read between the lines to find out what is being undertaken and from what data base.

(AVERY HILL)

Use sub-headings

Make sure each part of your write-up is clearly identified by using sub-headings. The headings given on page 41 are specified only for one syllabus but they offer a helpful model for you to follow even if your syllabus does not require it.

Apart from signalling to the reader where you are going next, such sub-headings also act as a check on yourself to make sure that you have given each part of the write-up enough attention. Too many projects are marked down because they over-emphasize one part of the work and omit or deal sketchily with others. For example, a lengthy account of what information is collected is unwise if it is at the expense of explaining your analysis of the data and the conclusions you have reached.

Referencing your sources

Where you get your information from – whether it was from books, maps or talking to people – should be clearly acknowledged. You will be given credit if a wide range of sources is seen to have been used and, at the other extreme, the marker will certainly notice if part of your write-up has been copied without acknowledgement from a standard book on the subject.

When presenting your results, it is not for your teacher when marking the project to find a way through the information you have supplied; it is for you to guide him or her through. As one syllabus puts it:

Candidates should be encouraged to maintain a high level of work presentation. All sketches, diagrams, maps and graphs included must be referred to in the text. Maps should be clearly labelled to show title, compass direction, key and scale.

Bulky items, such as town guides, should be kept to a minimum. Any excess material of this nature is normally to be found in the appendix.

(SEG B)

Maps, diagrams, tables, photographs, etc., should have clear titles and be:

either **(i)** included in the text at the point you refer to them;

or **(ii)** collected at the end of the text with cross-referencing to each item whenever it is mentioned.

Neatness

There may or may not be marks allocated for presentation, but neat, legible writing and well-drawn maps and diagrams do inevitably affect the marker's impression of the quality of your work.

Example

No one example can illustrate the many possibilities of topic, techniques and approaches to geographical studies which are opened up by the fieldwork projects. These guidelines, prepared by a teacher for work initially undertaken by the whole class together, may give you some indication of what doing a group-based fieldwork project would involve:

'The landscape of the northern area of the Peak District National Park is related to its geology.'

This is a *hypothesis* that you are investigating. It sounds complicated but it is just a *statement* or *idea* about something which you can then *test* to see if it is true or not.

On your day out, from Chesterfield onwards, keep a look out for any evidence of the rock type affecting the landscape.

The rock types you will see will be mainly: Carboniferous Limestone
shale
Millstone Grit (a coarse sandstone)

These rocks are all sedimentary.

Know what each type of rock is like: how was it formed?
what colour is it?
how much of the area does it cover?
is it permeable or impermeable?
what special features would you expect to find?

(Are there other rock types in the area?)

What to look for beyond Chesterfield (roughly in the order of their appearance):

Field boundaries (dry stone walls)
Building materials – is it local stone?
PDNP sign – what is it?
Froggatt Edge
*Relief/the shape of the land (height and slope)
*Land use – crops and animals, for example
*Drainage . . . or lack of it!
Gorge
Dew ponds, baths
Dry valleys
Lorries . . . what sort in particular?
Bradwell Gorge
Cement works
Hope village, followed by Castleton
Treak Cliff cavern
Mam Tor – the mother mountain
Tourist shops; information; accommodation; entertainment
Edale, in the valley of the river Noe
National Park information centre

Start of the Pennine Way/footpath erosion
Scree
Landslips
Recreational activities
Sink-hole or swallow-hole (pot hole!)
Blue John cavern
Winnats Gorge
Scars
Vegetated scree slopes
Speedwell cavern
  *to be looked for in all locations

What do we mean by landscape?

Landscape means anything you can see in an area. This can be:

1 *Physical landscape* – the natural things such as relief (rock outcrops, hills, valleys), soil, natural vegetation, drainage.

2 *Man-made landscape* – the unnatural things such as farming (crops or pasture?), man-made vegetation, industry, settlements, tourist attractions.

You should look out for any *connections* between what you can see and the rock types. Make a note of all these connections – you will need them to write up your study.

Draw sketches of what you see. Make sure that you label them and note what they show so you can recognize them when you get back to school. By all means take photographs of significant features, but these should *supplement* and *not supplant* your field sketches. Buy postcards of illustrations of rock type and landscape. Ask your cavern guide questions.

Remember: how good your study is will largely depend upon how much information you record (make a note of) when you are *in* the Castleton area.

Writing up your study

The following is an outline of how you should organize your course study:

1 Introduction   Statement of hypothesis – explain what you are trying to prove or disprove.
Where is the northern area of the PDNP? Possible map.
Include natural and man-made landscape.

2 The geology of the area   Map.
Explain what the rock types are/where they occur, but ... do *not* spend too much time on the explanation of how each rock type was formed!

3 Data collected   This is the main body of your work, i.e., it should contain all the connections you can find between the rock type and landscape. Include any *interpretation* of what you have found.

This section is worth 9/15 marks.

4 Conclusions   What conclusions have you come to? Is the hypothesis proved or not?

(Nicholas Chamberlaine School, Bedworth)

SUMMARY

Perhaps the last word on projects should be left to an examiner emphasizing the worst and the best in the hundreds of projects he had reviewed when sampling the work sent in from different schools and colleges:

Many projects were obviously copied word perfectly from current geographical texts and showed no geographical value whatsoever.

Far too many candidates were attracted to glossy hand-outs, pre-packaged material, etc., which is of little use.

One can therefore define the best candidate as a person who has a flexible mind, who knows what technique to use to gather the information required, who knows how to illustrate the information, to analyse it and to evaluate it. The candidate knows how to relate it to the geographical background, lay it out with well-defined aims and objectives, and how to come to precise conclusions. This individual is one who also demonstrates pride in the work, and an enthusiasm in the preparation.

(WJEC)

OTHER TYPES OF COURSEWORK

The choice of work in the other two categories of coursework – coursework exercises and class tests – is a matter for your teacher. Unlike some individual or small-group projects, the question for you is therefore not how to make use of the scope which choice offers, but rather how to do well in tasks which are common to all GCSE Geography students in your school or college.

Yet while all your fellow students are doing similar work, there may be great differences between that work and what is done by candidates in other schools and colleges. Syllabuses typically allow teachers a wide choice of activities. For example:

Folio of work

(i) The folio must consist of two *separate* and *different* pieces of work covering a range of ideas and questions from either of the option modules.

(ii) Each of the pieces of work will be worth 11%. Each should take approximately 2 weeks (4 hours) classroom time.

(iii) *At least one* of the pieces of work must be selected from the following:

A structured assignment.

Practical test – e.g., in which a set of statistical data is converted into appropriate graphs, maps etc.

An extended essay – e.g., in which a local or national issue is analysed and opinions expressed.

The other piece of work may be based upon any appropriate technique e.g., an oral.

It is assumed that the folio will grow out of and be part of the normal teaching programme but must be directed towards the final assessment.

(Avery Hill)

Because the work will be particular to your school or college, it is not possible to offer guidance here on how to tackle it.

In one respect, however, the same point applies to these types of coursework as to projects. They are marked by your teacher and so, if you take the trouble, you can in many cases find out the basis on which they will be marked. For example, the decision-making exercises referred to on page 43 are marked according to this scheme:

LEVEL OF RESPONSE

Describing the problems

Max marks for each level

(i) The candidate has demonstrated a basic understanding of some of the processes which relate to the problem being considered and selected some of the data which is relevant to the question/s or hypotheses posed.   *0–2*

(ii) The candidate has demonstrated an understanding of a number of processes and/or concepts which relate to the problem being considered and selected data which is relevant to the question/s or hypotheses posed.   *3*

(iii) The candidate has demonstrated a thorough understanding of a range of processes and/or concepts which relate to the problem being considered and selected relevant data to support responses to the question/s or hypotheses posed.   4

Solutions

(i) The candidate has selected or suggested a feasible resolution of the issue.   0–4

(ii) The candidate has suggested and justified a feasible resolution of the issue, whilst recognizing one or two drawbacks or weaknesses in the proposed solution.   5–6

(iii) The candidate has suggested, analysed and commented upon different ways of resolving the issue, including showing an awareness of the limitations and value judgements inherent in the proposed solution.   7–8

Effects and values

(i) The candidate has suggested a possible effect of the proposed solution and appreciated the different attitudes to the issue that groups or individuals may hold.   0–4

(ii) The candidate has predicted possible effects of the proposed solution and appreciated that the different attitudes of groups and individuals to an issue reflect differing values, experiences and priorities.   5–6

(iii) The candidate has predicted a number of possible effects if proposed solution were to be adopted; analysed ways in which the values, experiences and priorities of groups and individuals may affect their attitudes to an issue and explained how the definition of, and response to, a problem may be influenced by the values and the attitudes of decision-makers.   7–8

(MEG C)

So, brilliant though your solutions to the problem of siting an industrial park on the edge of your town may be, you will only be eligible for 8 of the marks (those for solutions) available for the exercise (a total of 20) unless you also identify the main feature of the problem and predict possible consequences of the preferred solution.

As with all coursework, the opportunity is there for you to demonstrate abilities largely unrecognized within the confines of an exam room. Don't waste that opportunity by misjudging what is expected of you or what you will be given credit for.

SECTION 8
Understanding the examination papers

Whatever assessed coursework you have done during the year, your grade is still likely to depend mainly on the answers you give to questions in the exams at the end of the course. All GCSE Geography courses culminate in one or two unseen exam papers, the marks for which contribute between 40% and 80% of the total assessment. Doing yourself justice in those exam papers is clearly partly a matter of marshalling what you know, understand and can do at the required moment. Section 10 of this book offers some suggestions on revision.

But being prepared for the exam paper is also a matter of anticipating what to expect and knowing how to respond appropriately. In particular, before you arrive in the exam room you should know:

(i) what type of exam paper to expect;
(ii) how to interpret exam questions;
(iii) how to answer in a way which will gain credit.

After thorough revision of your course, you are half-way to being ready for the exam. An awareness of what will be expected of you will take you the rest of the way. It may also give you some confidence with which to tackle the exam papers. This section and the next will help you with that preparation.

MAKING SENSE OF THE EXAM PAPERS

When you reach the moment of a first look at an exam paper, some questions should please you, others may worry you. But nothing, apart from the particular wording of the questions, should come as a surprise. In setting the questions, the examiner has been following prescribed 'rules' described in the syllabus about such limitations as the topics to be included, the extent of choice (if any), the style of questions and the pattern of difficulty within the paper.

61

You could take a look at those 'rules' for yourself. Better still, familiarize yourself with how they are applied by studying previous exam papers on the same syllabus (or, with a new syllabus, the specimen papers which the examining group provides).

Use this list to check that you know what to expect in the papers you will be faced with:

1 Time allowed
Usually this is given as a block of time during which you must read the paper as well as prepare and complete your answers. Some exams do allow you a few minutes 'reading time' over and above the one to two hours available for answering the questions.

2 Number of questions set
This is usually stipulated in advance although, with most questions sub-divided into several parts, the number of question parts may vary.

3 Choice of questions?
One trend in Geography exams is towards candidates having to attempt all the questions set. This means you cannot afford to have weak or underprepared sections of your course because you will not be able to avoid attempting an answer on such a section. Though lack of choice may cause you some problems, papers where there is no choice should be fairer because every candidate is answering the same questions.

Even if there is no choice on one of your exam papers, there may be some choice on a second paper which either examines a different part of your syllabus or the same syllabus content but in a different way. Check how much choice you have and what rules govern the pattern of choice. For example:

PAPER 2/WRITTEN PAPER/2 HOURS/50% OF THE TOTAL MARKS

This will consist of *nine* structured questions from which candidates will be required to answer *four*, as detailed below.

Two questions will be set on Unit A – Population and Settlement: *one* question on Population (Key Ideas 1 and 2) and *one* question on Settlement (Key Ideas 3 and 4). Candidates will be required to answer *one* question from this Unit.

Two questions will be set on Unit B – Resources and Economic Activities: *one* question on Agriculture (Key Ideas 1 and 2) and *one* question on Industry (Key Ideas 3 and 4). Candidates will be required to answer *one* question from this Unit.

Three questions will be set on Unit C – Physical Environment and Human Activities (one question on each Key Idea). Candidates will be required to answer *one* question from this Unit.

Two questions will be set on Unit **D** – Development (one question on each **Key Idea**). Candidates will be required to answer *one* question from this Unit.

(SEG A)

4 *Type of question*

The structured type of question referred to above is used in every GCSE Geography examination. The examination papers for many syllabuses consist entirely of structured questions. Most attention in this section and the next is therefore given to interpreting and answering questions of that type.

There are three other types of questions used in certain examinations:

(i)   short answer questions (LEAG A; LEAG D; NEA B);
(ii) objective questions (LEAG A; NEA A);
(iii) problem-solving exercise (Avery Hill).

How to 'read' each type of question is explained later in this section. For the moment, just make sure you know how many of each type you will encounter in the examination.

5 *Variation in difficulty?*

When setting the questions, an examiner will have in mind some pattern of difficulty, either a mix of 'easy' and 'hard' questions or perhaps questions of 'equal' difficulty. That may make little difference to you since you may find easy what he/she thinks difficult or vice versa. But it is worth knowing, partly for allocating your time but also to keep your spirits up, if, for example, you can expect the final part of each question to be quite hard. Such a question would have been designed with a so-called 'incline of difficulty' (the 'easiest' part first and the 'hardest' part last). Don't despair; come back to the part you find difficult and in the meantime start afresh with the next question.

All these features will be spelled out in a statement such as this:

Paper *1* **(1 hour 10 mins.) (70 marks are allocated to this paper)**

Candidates will be required to answer all the questions which will be on mainstream topics. The questions, based on data, will be structured and have differing levels and inclines of difficulty. They will test knowledge, understanding, and a limited range of skills across the seven areas of study.

Paper *2* **(1¾ hours) 90 marks are allocated to this paper)**

Candidates will be required to answer 3 of the questions set, which will follow the order of the subject areas listed in the syllabus. A minimum of 7 questions will be set so that each subject area will be substantially

represented. The longer structured questions will have comparable inclines of difficulty and will test knowledge, understanding, a wider range of skills than Paper 1 and the application of values.

In either paper, an OS map may be used, and if used will include a key. Atlases must not be made available or used in the written examinations.

(NEA D)

Check the 'rules' for your syllabus.

MAKING SENSE OF THE QUESTIONS

Before deciding whether to opt for a particular question (if there is a choice) and how to answer it, a clear view of what it is asking from you is essential. Under the pressure of exam conditions it is all too easy to make the wrong choice and/or to answer the question you *think* is there rather than one which careful reading of the exam paper would show was the one actually set. Answering an exam question can be challenging enough without the extra, self-imposed handicap of having misread the question.

It may help if you recognize from the start that the examiner is trying to set questions which are on the syllabus, are clear in what they ask for and can be answered by everyone who has followed the course and has prepared adequately for the examination. It is obviously not in *your* interests if a question topic seems obscure or the question wording reads ambiguously, but it is not in the *examiner's* interest either. Questions which few candidates can answer or many misinterpret are unsatisfactory for the examiner. Checks are therefore made with a view to eliminating irrelevance, wordiness, ambiguity or undue difficulty in a question before the final version is printed and reaches your eyes.

The main features of each type of question can be readily recognized.

STRUCTURED QUESTIONS

A structured question is on a single theme but sub-divided into several parts, usually incorporating relevant items of information to which some or all of the parts refer. A great range of types of data is used in this way, including maps, diagrams, graphs, statistics, photographs and newspaper extracts. Although the parts are linked by a theme, answering the later parts does not usually depend on success in answering earlier parts.

64

Even a quick reading of a typical structured question, with perhaps a dozen parts and several pieces of data, can take some time. A *careful* reading, sufficient for you to be sure what is required, will take a bit longer. If there is no choice on the paper, you can read the parts separately and treat them as linked but distinct questions. If there is a choice, read through the whole question before deciding whether to answer it. The first part might seem inviting but the later parts are likely to be worth more marks and they may not be as well suited to what you can do. Because of their length and the variety of demands they make on you, identifying what is required is more difficult than for the other question types. Don't miss out by overlooking or misreading the signals given by the examiner. Watch out for two types of signal:

the exact words used in the question;
clues to the length of answer.

A *The exact words used in the question*
You will find it useful to get into the habit of recognizing the key words in a question, 'homing in' on what it is really about. Every question asks you to *do* something on a specified *topic* but there may be *conditions* which limit the topic or what you should do with it. So ask yourself:

1 What exactly is the topic?
One of the commonest errors of exam candidates is to write about a topic similar to, but distinct from, the one mentioned in the question. If you have conscientiously revised your notes on shopping patterns within a town you might be tempted to overlook, or simply not see, that the question is about *out-of-town* shopping centres. You should resist that temptation unless you can adapt what you know to make it relevant to the actual question. The marking scheme which the examiner will use takes account of the exact words used in the questions. If you, accidentally or deliberately, write on a different topic, the examiner will put a line through the answer and give it no marks. It is not for the examiner to search through what you have written in the hope of finding something relevant. It is for you to write on the topic set and only on that topic.

2 What exactly do I do?
The answer would be simple if you could assume that, irrespective of whether the question asked you to *describe, give reasons for* or *analyse*, you could read it as meaning *write all you know about*. Of course this is not so. The people who write the exam questions deliberately vary the demands they make on you, even within the same topic. Some of the instructions given – *state, name, list, calculate* – are plain enough, but

are you sure you understand the difference between, for example, *describe, compare, suggest the extent to which* and *discuss*? Each word or phrase is distinct in its demands and you should understand the differences. Working through the examples later in this section will help you be sure not only about the question topic but also what you are expected to do.

3 Are there conditions to look out for?
There shouldn't be hidden 'trip wires' in the questions because the examiner is not trying to trick you. But you can easily miss a word or phrase which qualifies the scope of the question or the form of your answer. It may be a single word such as 'name two *contrasting* examples' or a phrase such as 'use a simple sketch map to illustrate your answer'. Such conditions, though an integral part of the question, can easily be forgotten as you concentrate on the main thrust of your answer.

B Length of answer clues
You may be so perplexed by a question that you are hard-pressed even to find a few, preferably relevant, words for your answer. Alternatively, the examiner may have been kind enough to hit upon a topic you can tackle with ease (check: is it actually *that* topic or one which is superficially similar but significantly different?). If you are fortunate, you may be ready to write at length. Given the knowledge and the time, many of the questions could prompt an essay-length answer. But even if you have that knowledge, you cannot spare the time to write at length about something in an exam where there may be fifty or more part-questions to answer. Look for the clues as to how long to spend on each answer and how much to write:

1 Word signals
Where a question 'state the percentage of . . . ' is followed by 'state fully one piece of evidence for . . . ', the 'fully' is not there by accident.

2 Lines allowed
If the space provided in the answer book is not enough, you are allowed to continue on to a separate answer sheet. But, assuming you have average-size handwriting, the space provided is a strong hint as to the length of answer expected.

3 Marks allocated
Indicated in the margin of the question paper is the mark (or marks) for each part of the question - another good guide to the time you should spend on it.

A relatively straightforward eight-part question will illustrate how these checks can be applied when first reading an exam question:

2 (a) Study the map below.

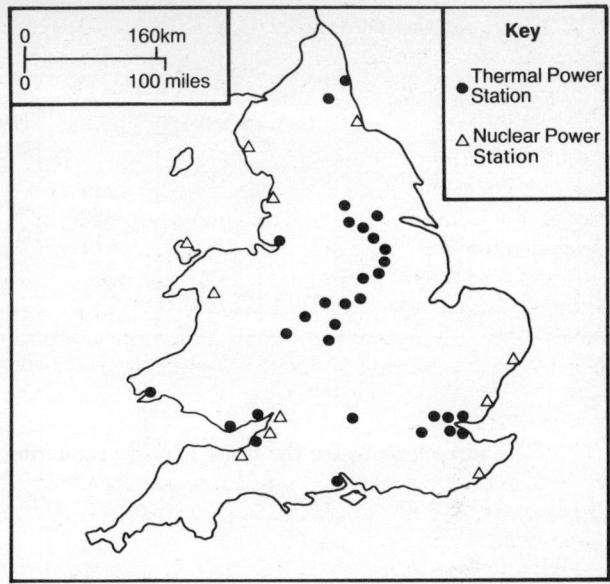

(i) Briefly describe the location of the nuclear power stations.

. . . .

(1 mark)

(ii) Give *one* reason for the location you have described.

. . . .

(1 mark)

(iii) Name the *two* main fuels used in the thermal power stations.

. . . .

(2 marks)

(iv) Give *two* arguments for, and *two* against, an increase in the number of nuclear power stations.

. . . .

. . . .

. . . .

. . . .

. . . .

. . . .

. . . .

. . . .

(8 marks)

(b) Study the table below.

Year	Production of coal in millions of tonnes	Manpower (in thousands)	Collieries	Output per man-shift (in tonnes)
1952	211	705	880	1.2
1957	207	708	822	1.3
1962	188	536	616	1.6
1967	164	410	443	1.8
1971	142	286	292	2.2
1975	124	246	246	2.2

(i) Give *three* reasons for the trend in coal production.

. . . .
. . . .
. . . .

(3 marks)

(ii) State *one* problem caused by the decline in manpower.

. . . .

(1 mark)

(iii) Give *two* reasons for the increase in output per man shift.

. . . .
. . . .

(2 marks)

(iv) There has been much debate about plans to develop new coalfields, for example in the Vale of Belvoir, an area of attractive farmland in the Midlands. Discuss the arguments for and against the development of such a coalfield.

. . . .
. . . .
. . . .
. . . .
. . . .
. . . .
. . . .
. . . .

(6 marks)
(NEA B)

A *1 Topic*
The first seven questions are brief, direct and clear but what is the topic of *(b)* **(iv)**? Are you to write about a specific area, the Vale of Belvoir, or are other UK coalfield developments, such as Selby, relevant?

A *2 Action required*
In the first seven parts, what you have to do is clearly indicated by the words at the beginning of each sentence, such as *briefly describe* or *state one problem*. It is *(b)* **(iv)** which again is less obvious. What do the vital words *discuss the arguments* (which don't occur until the second sentence) actually mean? Is it a different way of asking you to do the same thing as in *(a)* **(iv)**? No - *(a)* **(iv)** asked you to *give* the arguments; this asks you to *discuss* the arguments. In other words, you only had to recall and restate two arguments for and two arguments against nuclear power stations in *(a)* **(iv)**, but here you are required to discuss - to express an opinion about the merits of the arguments.

A *3 Conditions*
Watch for the *number* of points you are asked to make. For example, four arguments in all must be given in *(a)* **(iv)**. Also, don't overlook the point that the discussion in *(b)* **(iv)** should consider arguments both *for* and *against* coalfield developments.

B *1 Word signals*
Note the 'briefly' in *(a)* **(i)**.

B *2 Lines allowed*
The brevity of the description expected in *(a)* **(i)** is reinforced by only one line being given in the answer book for your answer.

B *3 Marks allocated*
Note that 14 of the 24 marks available are for just two of the eight parts. But don't go overboard with your discussion in *(b)* **(iv)**. It is worth only 6 marks compared with the 8 which merely giving four arguments would bring in *(a)* **(iv)**.

Detailed dissection of a question is not practical given the time available in the examination room. But what you can do is practise picking up signals such as these on first reading of a question. Once you are in the habit of checking for the signals, it can be done very quickly. You might try underlining the key words as you read through the question so that the exact instruction, or any partially concealed conditions, are not forgotten when you come to answer. The time spent on doing this is far from being wasted if it prevents you from going off-track in your answers.

Try applying the checklist used here to a reading of the question given at the beginning of this book (pages 1–3) and then to the questions which follow. In each case, some comments are given as to what you should be on the look-out for, but see if you can pick up the signals for yourself before looking at those comments.

Example 1

Refer back to the question at the beginning of the book (pages 1–3). What signals are there to be picked up in the eleven parts and twenty separate responses required by that question? Note these points:

(a) **(i)** Make sure you measure the 1982 column.

(b) **(iii)** Don't just list the changes; explanation is called for.

(b) **(iv)** Four marks for this; making one point will not be enough to earn you all 4 marks.

(c) Note that the scale of treatment of the topic moves 'down' here from UK regions to variations at the 'local' scale within the regions.

(d) **(i)** You coped with reading figures from bar graphs for *(a)* **(ii)**; now is the time to show you can do the same for triangular graphs.

(d) **(ii)** Definitions time; you should be prepared with answers to questions such as this. Don't forget to give the examples.

(d) **(iii)** After numerous, briefly worded, direct questions, here is one you may need to read twice. The *topic* is the relative importance of primary and tertiary industry in 'least' developed and 'highly' developed countries. *Action required*? *To what extent* invites a judgement about how far the statement in the first sentence is true. Just to accept the statement and give examples would not meet that requirement. The *conditions* attached are that you should relate your answer to Diagram 3 as well as to studies you have undertaken. You are provided with twenty lines in the answer book with a maximum reward of 12 marks.

Example 2

Some structured questions are less varied and substantially shorter than that example.

8 **Study Figure 14, Mount St Helens.**

Figure 14 Mount St Helens 18 May 1980

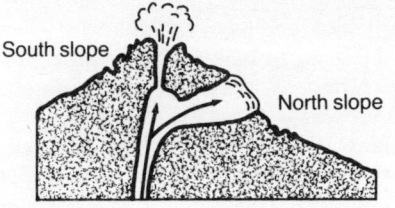

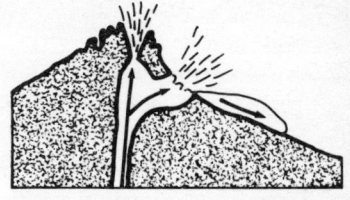

A 8.30 a.m.

Small ash and steam eruptions were rising from the crater. There was a large bulge on the northern slope. It was growing bigger as new magma, rising in the volcano's vent, was blocked by a thick crust of old magma.

B 8.32 a.m.

An earthquake (Richter scale 5.0) caused a great landslide of rock, soil, snow and ice from the bulging north slope.

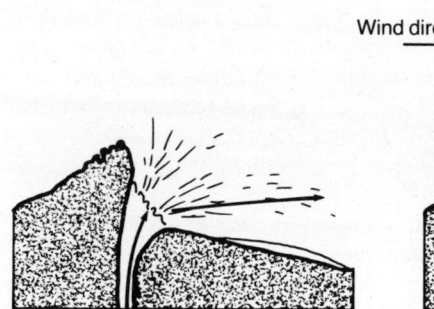

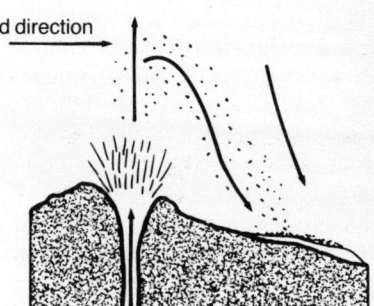

C 8.33 a.m.

The landslide allowed the magma in the vent to escape. There was a gigantic blast of gas, steam, dust and rock.

D later in the morning

Eruptions of gas and ash during the morning rose more than 20 km into the air before being deposited over a wide area.

(a) Mount St Helens is on the edge of the Pacific Plate. This is a sub-duction zone. What does this mean is happening to the plate? *(1)*

(b) Describe how the shape of the mountain changed from 8.30 to 8.33. *(1)*

(c) The magma was highly viscous (thick and pasty). This meant gases could not easily escape. They built up to high pressure and explo-sive eruptions resulted. Some other volcanoes have low viscous magma (thin and runny) which escapes without explosions.
Suggest how the events of 18 May 1980 would have been different if the magma had been of low viscosity (thin and runny). *(1)*

(d) From which direction was the wind blowing? *(1)*

(e) What effect did this have? *(1)*

(MEG A)

This example is from the 'easiest' of the three specimen papers for the MEG A syllabus (with answer lines deleted). If you took that paper you could expect a grade within the range D to G. Each part of the question requires a brief answer of no more than a sentence in length and perhaps only a word. Even in a short question of this type, there are some points to note if you are to earn the full 5 marks available:

(a) Refers back to your study of different types of plate boundary. Do you remember what happens in subduction zones?

(b) Unlike *(a)*, this can be answered by describing what the diagrams show.

(c) Asks you to link what you know about what happens to low viscosity magma to this specific example.

Example 3

Relatively brief questions are not necessarily the most straightforward ones. Here is a question, structured in only three parts, from the 'most difficult' of the papers set on the same syllabus as the question on Mount St Helens.

5 (a) Explain why there is a mismatch between the availability of water and the demand for water in the areas shown on Fig.4.

(10)

(b) The map outlines a possible solution to the problem.
   (i) What environmental risks are associated with the planned scheme? *(6)*
   (ii) Would the construction of barrages on the Wash and Morecambe Bay offer a better solution? *(4)*

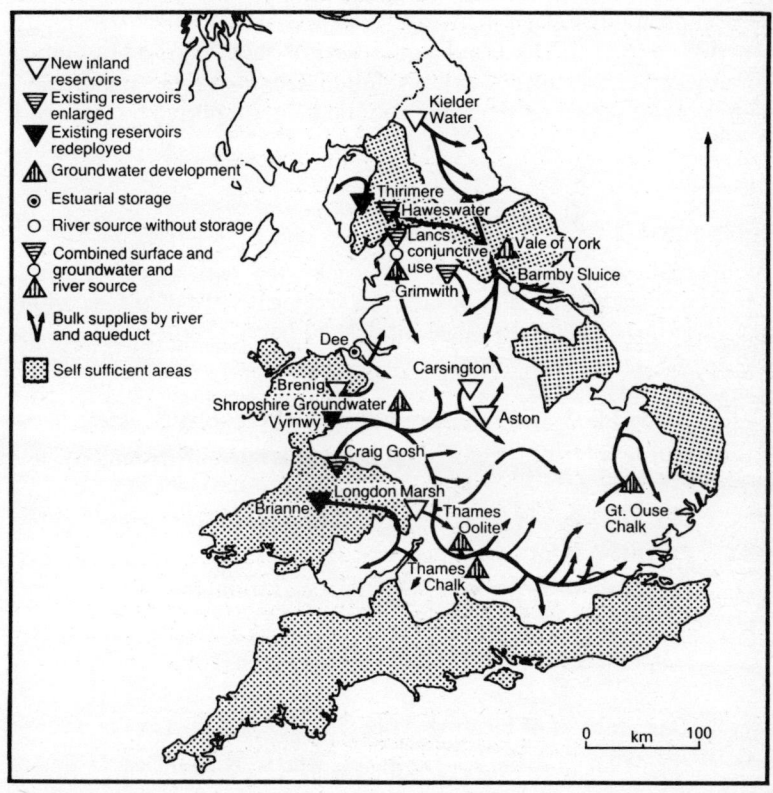

Fig. 4 A £1,500 M strategy for water

(MEG A)

If you are entered for this paper you will become familiar with this type of question and the longer answers – amounting to short essays in some cases – expected of you. The questions do not demand point-by-point extraction of information from the map but rather an extended consideration of the topic to which the map refers. Given the form of the question paper, with answers written in a separate answer book, the mark allocation is the only clue as to how much you should write. Beyond that, a successful attempt at an answer depends on correctly interpreting the three different aspects of the topic:

(a) Describing the 'mismatch' between areas of water surplus and the areas of highest demand is not enough. The question asks *why* it is so.

(b) (i) Before tackling this question, are you sure you know what 'environmental risks' are?

(b) (ii) Be careful. This is not a question just about barrages but rather about whether they offer a better solution than the supply and distribution network shown on the map. Your opinion, supported by argument, is called for.

Example 4

With syllabuses where all candidates take the same paper, you can expect a range of difficulty of task and varying lengths of answer within the same question (answer lines all deleted here). (Photograph B is not reproduced here; Figs. 10 and 11 are on page 76.)

6 *(a)* **Study Fig. 9 and Photograph B in the Resources Booklet.**

    (i) **In which part of central London is most office employment located?**

    (ii) **State *two* features of the area where most office employment is located.**

    (iii) **State *three* reasons why most office employment is located in that area.**

    (iv) **Why are most offices in cities in tower blocks?** *(8)*

  (b) **One problem of locating offices in cities is shown in Fig. 10.**

    (i) **What problem for the people who work in offices is illustrated in the cartoon?**

    (ii) **What problem for the city transport system is illustrated in the cartoon?**

    (iii) **State *two* other difficulties which are faced by offices which are located in cities.** *(5)*

(c) To overcome the problems of being located in cities like London many offices have moved to new locations.

Name one office complex which you have studied which has moved to a new location.
Describe the location to which the office complex has moved and explain why it was chosen. *(5)*

(d) Study Fig. 11 which shows the plan of a proposed office development in a large city. Suggest why this site is considered a suitable site for office development and what the feelings of those who occupy the neighbouring buildings would be, giving reasons for your answers. *(7)*

Total 25 marks

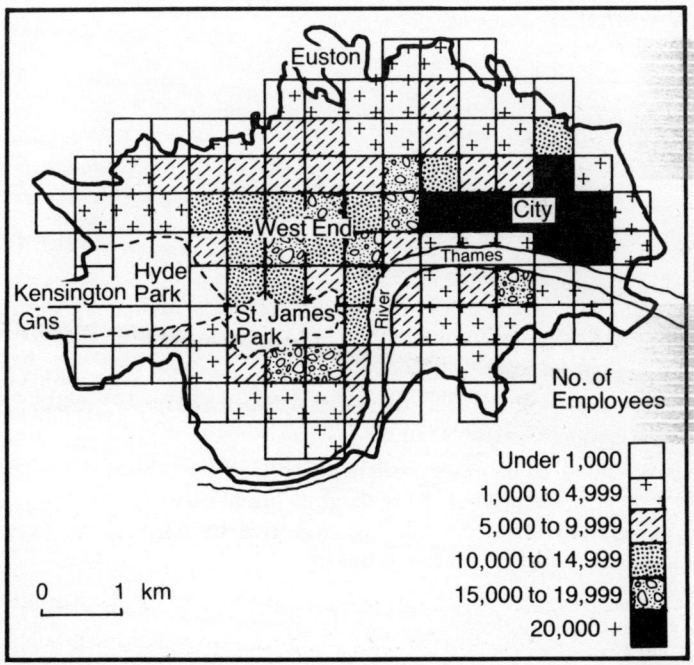

Fig. 9 Employment in offices in Central London (for use in Question 6)

**High commuting costs
could mean the end of the line
for your London office**

When fares go up, staff ask for more.
Or they leave and work locally. Do you really
need that London office?

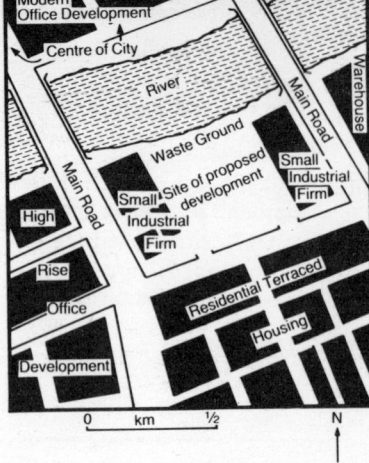

Fig. 10 (for use in Question 6)

*Fig. 11 Plan of area surrounding
proposed office development (for
use in Question 6)*

(LEAG B)

(a) and *(b)* are based on the four sources of information on the topic
given in the Resources Booklet (the photograph is not reproduced here).

(c) Moves on to look at the related topic of office relocation. Note
that there are three separate demands – *name, describe, explain* – to
each of which a response is required.

(d) Is intended to be more challenging still, each of the two demands –
suggest why this site, what the feelings of residents would be – requiring
you to relate your understanding of office location to an example
which you have not encountered before.

Example 5
Another example of a question with notably varied demands is taken
from the more difficult of the two papers set on syllabus MEG D.

8 Fig. 9 shows urban land use in Calgary, Canada. Below the map is
a brief description of the railways and roads running into the town.

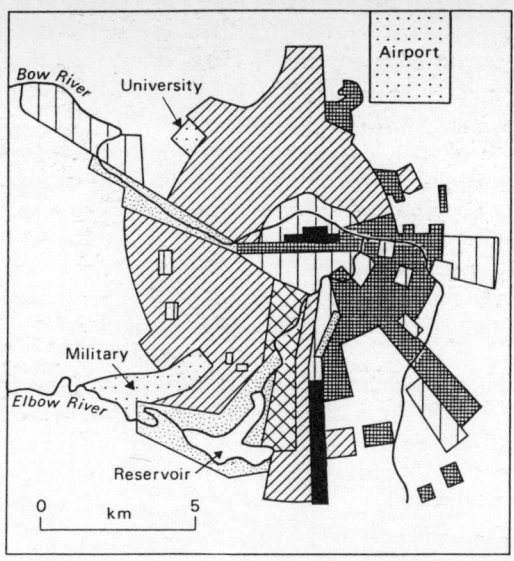

The Canadian Pacific Railway runs north west – south east through the town. Major roads run north to Edmonton, south to Lethbridge, west towards Kicking Horse Pass and east to Medicine Hat.

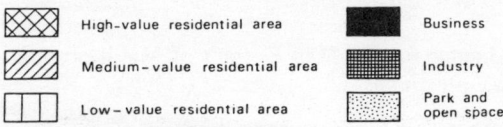

Fig. 9 Urban land use in Calgary

(MEG D)

(a) Describe the locations of the two business areas in the town. (2)

(b) Through which major land use does the road towards Edmonton pass? (1)

(c) Using the outline provided*, produce a sketch map to show where you think Calgary's main roads and railways are likely to run. Add notes to your map to explain your decisions.
*A blank outline of Calgary would be provided. (9)

(d) Compare the pattern of urban land use in Calgary with that of another town or city you have studied. (8)

(e) State, giving your reasons, in which part of the town of Calgary you feel it would be least pleasant to live. (5)

Although parts *(a)* and *(b)* ask no more of you than putting features of the map into words, parts *(c)* and *(d)* require considerably more than that and the bulk of the question's 25 marks are allocated to them.

(c) From your knowledge of other towns and cities (you would not have studied Calgary before), do you understand land use patterns sufficiently well to be able to insert where you would expect the major roads and railways to be in Calgary? If so, are your cartographic skills up to the task of completing *and annotating* the sketch map required?

(d) Do you know the land use of another town or city well enough to be able to *compare* the pattern there with that shown on the map? Note that the word 'compare' invites a point-by-point account of the similarities and differences. Two separate accounts do not comprise a comparison.

(e) Having completed that comparison, are you alert enough to recognize that part *(e)* asks for a more personal response? Where in Calgary would you least like to live? Don't ignore the signal that the marks will be allocated not for the wisdom of your choice but for the *reasons* (plural) which you offer.

Example 6

The variation of demands within a question is also well illustrated by this question (based on an Ordnance Survey map not reproduced here, and answer lines deleted):

2 **Use the Ordnance Survey Map Extract of Middlesbrough to answer the following questions.**

   (a) **A large chemical company has a works at grid reference 5324 on the map, and uses oil as one of its raw materials.**

   (i) **Give *two* forms of transport which are available for use by the works.** *(1)*

   (ii) **Give *three* advantages, other than transport, of the site of the chemical works.** *(3)*

   (iii) **The company also has a large site at Wilton on grid reference 5722 and another on the River Tees at 4722.**
   What advantages are there for a company having the main parts of its works close together? *(2)*

   (iv) **A decision has been taken to build another chemical works like the one at 5324. At first *three* sites were suggested in squares 5720, 5717 and 5825.**
   Which of these sites would you recommend the company to choose? Give the reasons for your choice.
   Reference of site chosen: ———— *(4)*

(v) The company finally decided to reclaim marshland and build a works at 5325. There were objections to this proposal from naturalists who suggested the company should build elsewhere as the area is used by large numbers of migrating birds.

How would you answer these objections on behalf of the company? *(4)*

(b) Much of the industry in the area shown on the map extract is declining. The area has been designated a Government Development Area. Using evidence from the Ordnance Survey Map Extract and the Map, Figure 3, explain what advantages the local council could advertise to attract new light industry to the area. *(6)*

(Total marks 20)

(NEA C)

(a) **(i)** Depends on your being able to use a grid reference and, having found the location, to read the map symbols correctly.

(a) **(ii)** Requires more than map reading; some interpretation of map evidence is needed to suggest the advantages the site possesses. Note the condition – *other than transport*.

(a) **(iii)** Refers to the map but draws on your wider understanding of industrial location.

(a) **(iv)** Asks you to make a choice and then argue the case for it. A description of your chosen site is not enough; you must justify it against the other possibilities.

(a) **(v)** Invites you to put yourself in the position of the company and express their point of view. In Example 4 part *(d)* you had to imagine yourself in the position of residents faced with the prospect of office development in their neighbourhood. Here you are asked to put yourself in the position of the company which wants to build a chemical works in spite of naturalists' objections. One important feature of GCSE Geography is that whatever the subject of study, you should be able to imagine the feelings and points of view of the people you are studying. This is called 'empathy'.

(b) Is notable partly for its complexity. The key word, *explain*, does not occur until towards the end of the question. What you have to explain is very specific – not just the general advantages of an area for industrial development but those advantages which a council could advertise to attract new *light* industry. Relevant information is given in the first two sentences, while the last sentence also includes the condition that you should use evidence from the two maps provided in your answer.

STRUCTURED QUESTIONS: A SUMMARY

The main points to note about structured questions are:

1 *They require you to be able to interpret a wide variety of data sources. Even the few examples reproduced here have referred to maps and graphs of various types, statistics, diagrams, photographs and cartoons. You can also expect to have to interpret other sources such as newspaper extracts and advertisements.*

2 *They make demands on you of many different kinds, from remembering a single fact to discussing an issue. The best evidence on what demands you can expect comes from past papers. Ignoring the content of the questions, check what past or specimen papers asked for. Some demands are specific, such as reading information from graphs, calculating percentages or recognizing map symbols. Others call for quite different skills such as presenting an argument or writing about what you think the feelings of a certain group of people would be.*

3 *To make sense of the questions, you must be able to recognize the differences in the instructions given. Some of these – name, state, describe, explain, give reasons for – are relatively easy to interpret once recognized. But are you sure you know what is expected from some of the other instructions which have occurred in these examples – compare, discuss, to what extent? Other potentially awkward demands in the examiner's vocabulary include words such as analyse (examine the elements of) and assess (usually meaning to estimate the quality of).*

4 *Structured questions require answers varying from a single word to an extended argument. There are verbal and non-verbal clues to the type and length of answer expected.*

SHORT-ANSWER QUESTIONS

Such questions are brief and direct. They can be explained with equal brevity and directness.

Even under exam conditions, what the question is asking for should be clear to you:

B 32 The table below shows population figures for five countries in 1980

Country	Area in '000s km^2	Population in '000s approx.	Density per km^2
ARGENTINA	2 800	24 000	
NETHERLANDS	30	13 600	453
NORWAY	320	4 000	13
SOUTH KOREA	100	33 000	
USA	9 400	212 000	23

(a) **Work out, and then write in the table, the population density of Argentina and South Korea to the nearest whole number.** *(1)*

(b) **Which of the five countries is the most densely populated?** *(1)*
 (LEAG A)

Where the questions are grouped together in a series, there is nothing to distinguish what a syllabus calls short-answer questions from the structured questions explained above. For example:

4 **Study the bar graph below giving information about delegates at Blackpool conferences.**

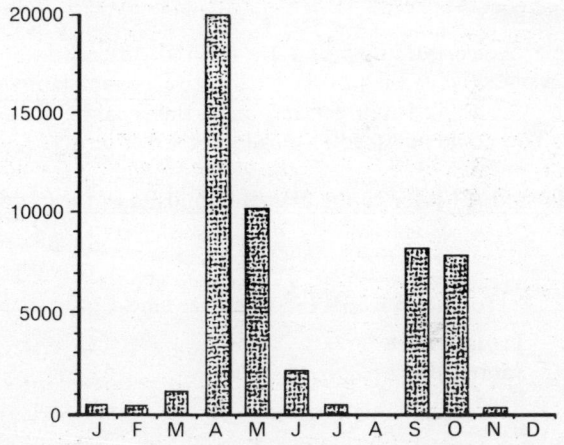

Delegates at Blackpool Conferences

(a) Name the month in which there are most conference delegates visiting Blackpool. *(1 mark)*

(b) Give *one* reason why July and August are of little importance for conferences. *(1 mark)*

(c) With over a quarter of its workforce employed in the holiday industry, Blackpool has problems of unemployment at certain times. In which season is there likely to be the highest unemployment? *(1 mark)*

(d) State *two* measures (other than attracting conference delegates) taken by Blackpool to reduce seasonal unemployment.

(2 marks)
(NEA B)

OBJECTIVE QUESTIONS

This type of question allows the examiners to ask for a wide variety of answers in a short time because so little of your time is spent in writing. You may be asked, for example, to:

(a) remember facts, definitions etc.;
(b) recognize features on diagrams or maps;
(c) choose from alternative explanations;
(d) suggest locations;
(e) use skills of map and graph reading.

What you have to do is choose from the alternative answers offered to you in one of three different ways.

Multiple choice

With multiple choice questions you are offered, usually in a list but sometimes in other ways such as on a map, five possible answers. You choose which you think is correct and mark the separate answer sheet accordingly. You could be asked to recall the correct term:

Questions *A2* and *A3* refer to the diagram opposite of the hydrological cycle.

A2 **Which of the following words is correct for label P?**

A	Transpiration
B	Saturation
C	Evaporation
D	Run-off
E	Solution

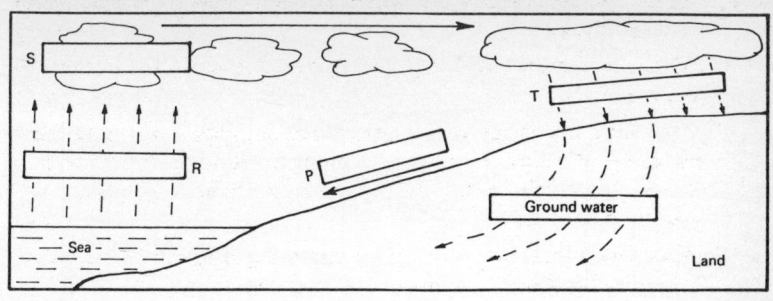

A3 **Which one of the following groups of words is correct for the labels R, S and T?**

	LABEL R	LABEL S	LABEL T
A	precipitation	condensation	evaporation
B	evaporation	precipitation	condensation
C	evaporation	condensation	precipitation
D	condensation	evaporation	precipitation
E	precipitation	evaporation	condensation

(LEAG A)

or to choose an explanation:

A4 **During the last twenty years many young people have moved from the rural areas of 'developing' countries to the large cities mainly because**

A	good housing is available for factory workers
B	tribal custom encourages them to move
C	governments encourage them to move
D	factories are short of workers
E	they think that well-paid jobs are available

(LEAG A)

or to interpret information on a graph:

A19 The graph below shows the birth and death rates for England and
Wales, 1700 – 1980.

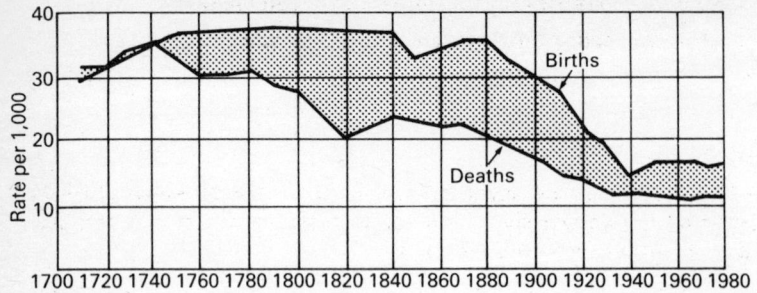

In which one of the following years was there the highest rate of
natural increase in population?

    A   1723
    B   1780
    C   1818
    D   1885
    E   1970

(LEAG A)

Matching pairs

With this type of question you have to match a statement to one of the
five possibilities offered:

In Questions *29–33* each group of questions has a set of responses, A, B,
C, D and E. In each group each letter may be used once or more than
once. For each question select the best response and mark its letter on
the answer sheet.

(NEA A)

For example:

Questions *31–33*

    A   **large-scale resettlement of refugees**
    B   **unrestricted movement across frontiers by migrants**
    C   **daily migration to and from work**
    D   **a shortage of people to settle fully all habitable parts of the
           country**
    E   **large numbers of farmers leaving their land and migrating to
           the cities**

Which letter (A to E) indicates a state of affairs typical of

31 highly advanced countries such as the USA and Britain?

32 developing countries such as India, Brazil and Mexico?

33 underdeveloped countries such as Australia and Brazil?

(NEA A)

Multiple completion

With this type, there are again five possibilities, one of which must be recorded on the answer sheet, but the instructions are more complicated. On one paper they are as follows:

In each of the questions *34* to *50* one or more of the responses is/are correct. Decide which of the responses to the questions is/are correct and mark A, B, C, D or E on the answer sheet as follows.

A   **if (1) alone is correct.**
B   **if (1) and (2) only are correct.**
C   **if (1), (2) and (3) are correct.**
D   **if (2) and (3) only are correct.**
E   **if (3) alone is correct.**

(NEA A)

Like multiple choice questions, what you are asked to do may be to recall information from your course:

34 Characteristics of less developed countries include which of the following?

(1) a high rate of population growth

(2) a large percentage of the population employed in manufacturing

(3) a high consumption of electrical energy

(NEA A)

Or the answer may involve some interpretation of data supplied (question 35 refers to the population pyramid on the following page):

35 The population pyramid is typical of countries in which:

(1) birth rates are high

(2) death rates are high

(3) medical care is available to most

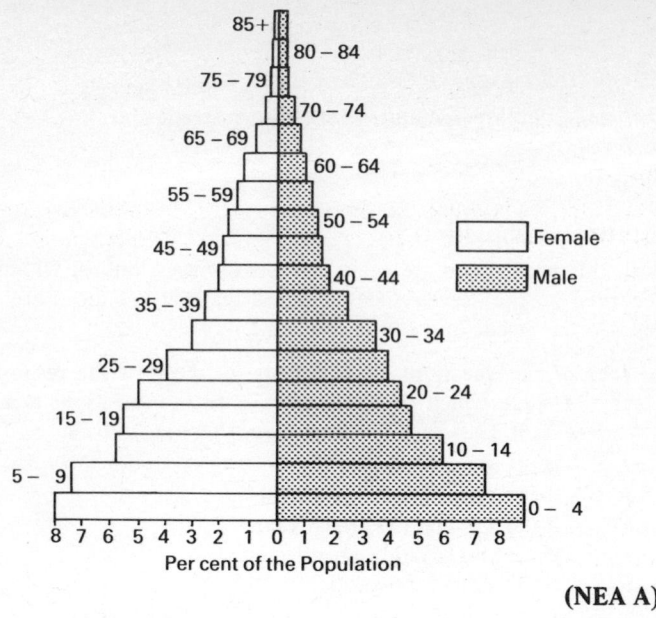

(NEA A)

All objective questions are so direct that there is little room for doubt as to the topic. Familiarity with what you must do to answer will come with practice on the type(s) of question(s) set on your syllabus.

PROBLEM-SOLVING EXERCISES

The main aim of studying geography must be to make use of what you learn by applying it to situations outside the classroom, both during your course and long after you have finished studying. Some types of coursework, especially problem-based projects and decision-making exercises, give you opportunities to apply your skills in real-life situations. Some structured questions in the final exam paper also, in a limited way, test whether you can use what you have learned to understand an unfamiliar situation. A problem-solving exercise takes one situation, supplies you with some information, and asks you to explain what *you* would do.

This example, a part of which is reproduced here, concerns a ring road around Bristol and is intended to:

(a) show why a road is needed;

(b) show what effects it will have when built;

(c) decide which particular route it should follow.

You are given an hour to complete the whole exercise. The third of four parts to the exercise looks like this:

Part C

Study the map below. It shows four possible routes for the northern part of the ring road. The problem is to decide which one to choose.

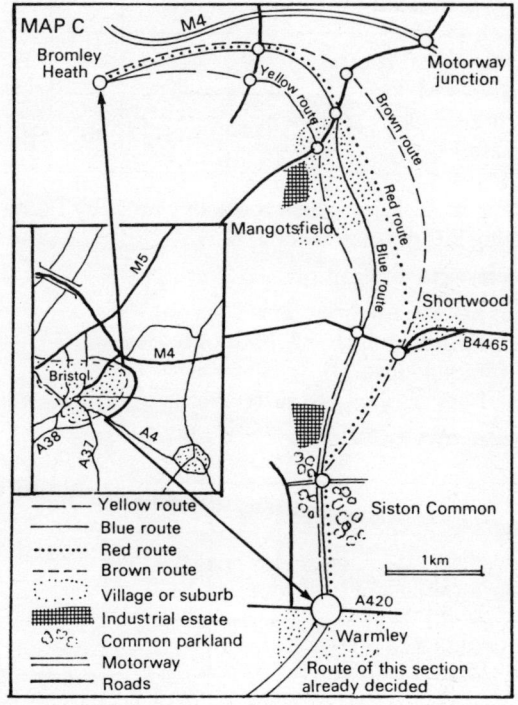

(i) **Which of the four routes:**

    1 **is the longest?** *(1)*

    2 **go through residential areas?** *(1)*

    3 **would serve the industrial estates in the area best?** *(1)*

(ii) **What is the length of the shortest of the four routes (from Bromley Heath to the A40)?** *(1)*

(iii) **From map evidence only, which route would you choose?**
Route: ————   *(1)*
Give *two* **good reasons for your choice.**   *(2)*

Here is more information about the four routes:

Table C	Yellow	Blue	Red	Brown
Cost (£ millions)	12.5	12.6	12.9	13.7
Number of houses to be demolished	15	16	11	12
Area of farmland lost (hectares)	19	24	23	24
Number of houses likely to suffer noise increase	147	96	104	106

(iv) **Go further in deciding which route to choose by filling in the table below using information from Table C.**

Rank each route in terms of:

Cost (this has already been done for you)
Number of houses which will have to be demolished.
Area of farmland lost.
Number of houses likely to suffer from noise.

In each case the lowest ranks 1 and the highest ranks 4

	Yellow	Blue	Red	Brown
Cost	1 (lowest)	2	3	4 (highest)
Properties demolished				
Farmland lost				
Noise				
TOTAL				

(4)

(v) **Which route seems to be the best from the results of this table?**
Route: ———— *(1)*
Explain why. *(2)*

(vi) **Do you think a simple ranking exercise is enough to base your decision on as to which route should be chosen?**
Say why. Yes/No *(3)*

(Avery Hill)

Part D introduces other features of the proposed four routes before you are asked finally to explain and justify (at some length) your preferred route.

In some respects, what you are asked to do in such an exercise is very similar to the demands which structured questions make. The difference is that a problem-solving exercise culminates in you using your geographical expertise to make a decision which you can justify.

SECTION 9

Answering examination questions

PLANNING YOUR APPROACH

When you open an exam paper, do you start immediately to answer the first question? If so, restrain yourself for just a few minutes to allow time for a little planning of what you are going to do.

Choosing which questions to answer

If there is a choice, you should already be familiar with the number of questions you have to answer and any rules about the pattern of choice (page 62).

Choosing between structured questions can be especially hazardous. You can easily be encouraged or deterred by the first thing about a question which catches your eye – perhaps a familiar map or a forbidding set of statistics. Try following these steps to guide you to a sensible decision.

1 Look at the two or three parts of a question which carry the most marks. Don't be swayed in your choice by a part which is worth only 1 or 2 marks.

2 Ask yourself both whether you are reasonably familiar with the topic and if you can do what it asks of you. Don't look only at the topic. You may find yourself more at ease with what you have to do on a question where the topic is initially uninviting.

3 Eliminate any questions on which, either because of unfamiliarity with the topic or because you don't like the task set, you will not perform well.

4 Make a mark against each of the questions you will answer.

A careful choice of this kind need only take five minutes or so at the beginning of the exam. If it ensures you can then settle down to answering the questions you are best placed to answer, it is time well spent.

Allocating your time

If, as is usually the case, the total marks available for each full question are the same, you should also plan to spend roughly an equal amount of time on each answer. You may well be able to schedule your answers before entering the exam room if you know in advance how many answers you have to write. Allowing (where there is a choice) a few minutes at the beginning for reading the paper and choosing questions, you can write down the time you expect to finish each of your answers. Keep that schedule in front of you. Don't panic if you are a few minutes behind schedule. But don't allow yourself to miss out on the 'easier' marks on your last question because you haven't left yourself any time in which to answer it.

Answering questions in sequence?

Most people prefer to answer questions in numerical sequence. It is certainly the plan least likely to result in accidental omission of questions. Even though generally answering in sequence, you may prefer to leave until later answering any parts of questions which prove difficult at first sight. If you do this, make sure you mark the omitted question clearly so that you don't have to search for it when time is running out at the end of the examination.

But you do not *have* to answer questions in the order in which they appear. You may prefer to tackle first the one which suits you best, to give a boost to your confidence. If you do plan your own route through the paper like this, be extra careful about clearly marking which questions you will answer and about keeping to your schedule.

HOW ANSWERS ARE MARKED

Having gone through all the preliminaries of learning how to interpret questions and planning your approach, the one remaining but crucial issue is how to answer in a way which will gain credit. Before studying some sample questions and how they should be answered, read the following insight into the thinking of an examiner in setting a pre-GCSE 16+ exam paper. It is reproduced here without the relevant data, but accompanied by an explanation of the examiner's thinking.

C6 Both the vegetation and climate illustrated on page 5 of the folder, CS/i/A, occur in the same region of the world.

 (a) (i) What name is given to the vegetation?
   (ii) What name is given to the climatic type? *(1)*

 (b) (i) Describe *four* features of the natural vegetation you have named in *(a)* (i). *(2)*
   (ii) State *three* ways in which the vegetation is adapted to the climate. *(3)*

 (c) Study the climate graph. What is
   (i) the highest monthly mean temperature
   (ii) the lowest monthly rainfall total
   (iii) the annual temperature range? *(1½)*

 (d) Describe *three* ways in which the climate at X on the vegetation profile differs from that at Y. *(3)*

 (e) (i) With the aid of a diagram, explain why temperatures are so high and why they vary so little from month to month. *(3½)*
   (ii) Why have September, October and November slightly higher temperatures than all but one other month? *(1)*

Rationale and objectives

This was intended as a straightforward question on one of the four regions specified in the syllabus. An effort was made to put a rather greater emphasis on natural vegetation than has customarily been the case.

(a) The aim here was to test whether candidates could identify via a sketch an example of natural vegetation rather than merely recall it as a response to a climatic term.
(b) (i) developed the answer to *(a)*. It gave some reward to those who accurately examined the sketch, but rewarded fully only those who had prepared properly for the examination.
   (ii) This took description a step further by seeking relationships between vegetation and climate. The higher ability level demanded by this question was evidenced by the lower marks obtained. Expected answers included tree height and search for light; large leaf size as aid to transpiration; drip tips and water shedding; variability of leaf shedding period etc.

(c) Was a fairly simple test of graph-reading skill.

(d) Introduced the idea of microclimate. The sketch provided allowed innate ability some scope, but to achieve full credit demanded thorough preparation.

(e) (i) Aimed at probing candidates' understanding of the relationship between the incidence of the sun's rays and temperature levels and their appreciation of the consequences of the uniformity in the length of night and day in equatorial areas. A diagram was obligatory for full marks.

(ii) Was to test further the skill of graph interpretation noting relationships between rainfall, cloud cover and insolation levels.

(from: *CS Geography: What's it all about?* (WJEC)

If you are well practised by now in the interpretation of questions, the examiner's 'rationale and objectives' should tell you little that you could not have gathered from a careful reading of the question.

When it comes to marking candidates' answers, an agreed marking scheme was used by every examiner who marked that paper:

Marking scheme

(a) (i) Selva etc. Rainforest. Eq. Trop.
(ii) Equatorial (tropical rainy). $2 \times \frac{1}{2}$ *1*

(b) (i) Three tiers, great height, extensive canopy, liana, species variety, hardwood etc. lack of undergrowth. $4 \times \frac{1}{2}$ *2*
(ii) Tall to reach light; drip tips to shed water; large leaves to aid transpiration; irregular leaf shedding $\frac{1}{2} + \frac{1}{2}$ since no seasonal rhythm. Buttress roots $\frac{1}{2}$ + dev $\frac{1}{2}$. Any three 3×1 *3*

(c) (i) 27°C (ii) 130mm±⁵ (iii) 2°C/25°-27°C
$3 \times \frac{1}{2}$ *1½*

(d) No sun, greater humidity, lower diurnal range, lack of wind/ Converse reasoning allowed. Any three 3×1 *3*

(e) (i) Sun high in sky/so heat concentrated/direct path through atmosphere/so less heat loss./ Little monthly variation because night = day/sun always high in sky.
Any five points @ $\frac{1}{2}$ $5 \times \frac{1}{2}$ *2½*

Credit appropriate diagram for *1 mark* 1×1 *1*
(ii) Less rain so less cloud, more sun. 1×1 *1*
―――
15

Here is a sample answer to that question, together with the examiner's comments on it:

(a) (i) What name is given to the vegetation?   *Selva* ✓

    (ii) What name is given to the climatic type?   *Equatorial* ✓   1

(b) (i) *There are high trees above 35 metres. Then the large bushes & trees 30m. Then the other bushes underneath 20m, which get less light. Then the undergrowth which usually consists of decaying leaves, nothing can grow because of so little light coming through.*   1½

   Good.

    (ii) *Through the dense forests the high, branching trees get most light so they grow also do the bushes, because the sun's rays are hot and there is plenty of moisture, due to rain which falls practically every day.*   1

   Misunderstood

(c) Study the climate graph. What is

    (i) the highest monthly mean temperature   *26°C* ✗

    (ii) the lowest monthly rainfall total   *130mm* ✓

    (iii) the annual temperature range?   *2°C* ✓   1

(d) *The vegetation at X does not get any sunlight because of the trees above, but it supports the roots e.g. the trees by keeping the soil together, when Y leaves fall they decay and re-fertilize the soil.*   1

   Fails to answer question.

(e) (i) *The temps. are so high because of the position of the sun in the sky it rises so high and then there is rain but because the sun is always high throughout the year then the climate is high. The equatorial regions are usually 6°N and 6°S of the equator.*   ½

   No Diagram.

    (ii) *Because then the rainfall is not so high and because the sun is directly above and so hotter.*   1

   Nearly there

   ⑦

Can you see how many marks were lost by failing to answer the questions as worded, especially in *(b)* (ii), *(d)* and *(e)* (i)? One mark was also lost in *(e)* (i) because the 'with the aid of a diagram' instruction was ignored. The overall result is that a candidate who knew quite a lot about the topic gained less than half of the marks available on the question.

More generally, the way in which the marking scheme is set out is worth noticing. Note in particular:

1 It itemizes the points which, if mentioned, would be credited. To minimize inconsistency among the many people marking the answers, *only* the points mentioned in the marking scheme can be credited.

2 Where, as in *(a)* above, a single word or phrase is required, the acceptable answers are listed.

3 Where, as in *(c)*, a description or explanation is required, a brief summary of the points to be credited is given. To be awarded the marks you would have to amplify those points by writing full sentences. You would also have to make the points in a way which refers directly to the question set.

4 Where, as in *(c)* (ii), there is measurement, the limits of accuracy (in this case plus or minus 5 mm) are stipulated.

5 Note that it is the *number* of relevant points made which matters. In *(b)* (i) and (ii) and in *(d)*, the number of points expected is given in the question. In *(e)* (i), however, any five points mentioned are worth a ½ mark each. If you had referred to only two factors in your explanation, you would, however fully developed and accurate your answer, be eligible for only two of those ½ marks.

Another specimen marking scheme – for the question printed at the beginning of the book (pages 1–3) and analysed in the last section – illustrates several of the same points. In this case guidelines are published to indicate in broad terms how GCSE answers will be assessed, including the category of objectives the examiners believe they are testing in each part of the question.

SYLLABUS B

Paper 1

Specimen marking guidelines

The marking guidelines being distributed are offered as a guide to teachers. They are not definitive but indicate in broad terms how candidates' answers will be assessed. A marking scheme used in the operational examination will have been subject to a standardization procedure. Mark allocations are shown under the four Objectives headings: K – Knowledge; U – Understanding; S – Skills; V – Values.

			Marks			
			K	U	S	V
(a)	(i)	Northern Ireland			1	
	(ii)	15 . . . 12			2	
(b)	(i)	Sharp rise in whole country . . . greater growth in South East			2	
	(ii)	Textiles/shipbuilding . . . electronics	2			
	(iii)	Competition . . . low wage-rate parts of world . . . over-production . . . in developed world	1	2		
	(iv)	Fear . . . by employers . . . to invest . . . danger from terrorism	1	2		1
(c)		Disguises . . . black spots . . . high unemployment . . . e.g., Consett . . . seasonal unemployment . . . e.g., seaside resorts/Blackpool . . . and employment high spots . . . e.g., Tunbridge Wells	2	2	2	
(d)	(i)	53 . . . 35 (allow ±2% but must total to 100!)			2	
	(ii)	Extract raw materials . . . directly from the earth . . . farming . . . provide a service . . . for community/other industries . . . school teacher	4	2		
	(iii)	Superficially true when seen on graph . . . Nigeria/etc. are 'less developed' . . . USA/UK are well-developed . . . Switzerland and Sweden have highest GNP . . . yet below 50% in tertiary . . . USA has high tech/automation . . . few people needed in primary and secondary industries . . . yet these very important in economy . . . Thailand/etc. may be regarded as highly developed intellectually . . . yet economically under-developed . . . one-product economies . . . may not show well . . . e.g., oil producers	3	2	3	4
		Total			40	

(SEG B)

Refer back to the question (page 1) to see how closely your expectation of what is credit-worthy matches that of the examiner. Note these points:

(b) **(ii)** The industries mentioned are only examples; other suitable examples would be credited.

(b) **(iii)** Only the bare bones of an answer are outlined in the mark scheme. You would have to make use of points such as these to write a more fully developed *explanation.*

(c) Three possibilities are given in the mark scheme; two of them, plus an example of each, are required.

(d) **(i)** Don't waste marks by inaccurate reading of the graph: limits of accuracy are stated.

(d) **(iii)** The question doesn't ask you to agree or disagree with the generalization in the first sentence. It asks you to express a view on the extent to which the statement is borne out by the facts. The mark scheme reinforces the importance of recognizing the key words '*to what extent*' by referring to some of the exceptions to the generalization about primary and tertiary industries. There can be no one way of answering such a question successfully, so the mark scheme merely suggests a kind of argument and example which should be credited. Some of the examples are to be taken from diagram 3 but you should also refer to evidence from your own studies. The marks in this case are thus not for a predetermined list of points but for the quality of the argument and examples *you* choose to present.

ANSWERING STRUCTURED QUESTIONS

With an understanding of how to read examination questions and how the marks are given, you should be able to anticipate how to answer questions successfully. To do so in practice of course you will also need to develop the skills, knowledge and understanding to meet the question demands. The examples which follow are taken from various syllabuses and may not be on topics with which you are familiar. But study them to check how accurately you are reading the questions and anticipating how they should be answered.

Example 7

The following question on international migration is taken from the more difficult of the two alternative papers on the MEG C syllabus. You might have studied the example of migration which the question uses, but answering the question successfully does not depend on previous knowledge of the Gulf States. It is intended to test your understanding of the patterns and processes of international migration.

Table 2

Basic statistics for the six member states of the Gulf Co-operation Council (listed in order of population totals)

Country	Population mid-1982	Area (000 sq km)	GNP per capita 1982 (US$)	Life expectancy at birth (years)	Estimates of foreign work-force
Saudi Arabia	10 000 000	2 150	16 000	56	40%
Kuwait	1 600 000	18	19 870	71	70%
United Arab Emirates	1 100 000	84	23 770	71	90%
Oman	1 100 000	300	6 090	52	70%
Bahrain	400 000	1	9 280	68	40%
Qatar	300 000	11	21 880	71	80%

Source: World Development Report, 1984 (World Bank)

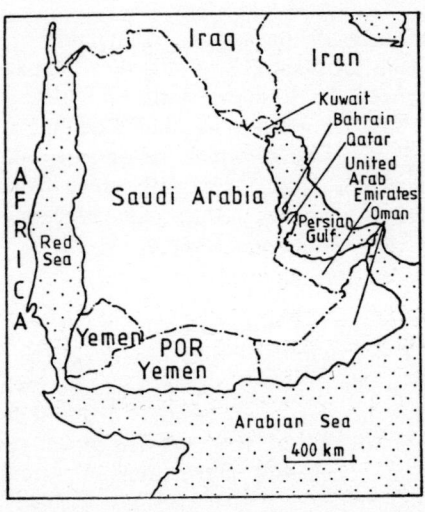

Fig. 4 Location of the six member states of the Gulf Co-operation Council.

(a) Study carefully all the information in Figure 4 and Table 2.

    (i)   Which state has the greatest number of foreign workers?   *(1)*

    (ii) The percentage of foreign workers in the six Gulf States in 1982 was 52%. Why might this average figure be misleading?

                                                        (3)

(b) Which of the six states would you consider to be the poorest? What factors led you to this decision?   *(2)*

(c) Read through the general summary below.

General summary of migrants in the Gulf States

The oil-rich Arab states have used their oil revenues to implement large-scale developments e.g., the construction of roads, hospitals, schools and oil refineries.

Their own populations are largely untrained in the skills required to complete these projects, so these states have had to attract workers from other countries. The Gulf States can offer much higher wages than many countries e.g., a job in Saudi Arabia may pay as much as six times the wage a Sudanese worker could earn at home for the same job.

In the early 1980s there were three main areas from which workers came:

    1 Egypt
    2 India, Pakistan, Bangladesh, Sri Lanka
    3 The Far East

The Egyptians have jobs in all the occupations from doctors to labourers. The workers from the Indian sub-continent are mostly in the service sector.

The workers from the Far East do much of the construction work, but also work in restaurants, hotels and other services (this is especially true of the Filipinos).

European workers are important in the Gulf States but they tend to occupy senior management positions, and these numbers are few in comparison with other groups mentioned above, e.g., in 1975 the numbers of migrant workers in the six Gulf States were:

Arabs	935 100
Asians	344 440
European	30 000
Others	77 800

Far-Eastern workers are now regarded as cheaper and politically safer than Arab or Indian workers as they are:

1 usually single;
2 working in remote desert areas;
3 do not have much contact with local people, whilst other groups have intermingled much more with local communities.

(i) Why do the Gulf States attract foreign workers in such large numbers? *(4)*

(ii) The summary suggests that the Gulf States prefer Asian workers. Why *should* the Gulf States have *any preference* as to the origin of their workforce? *(4)*

(d) Many European workers in the Gulf States work there for only a short period of time, a few months or possibly a few years. Why do so few European workers move to the Gulf States on a *permanent* basis? *(5)*

(e) For any named town or city consider the advantages and disadvantages of receiving a large number of newcomers. *(6)*

(Total 25 marks)
(MEG C)

Can you see how to answer each of the seven parts? Note these points:

(a) (i) The correct answer is Saudi Arabia (though the wording of the question and the table may cause some confusion).

(a) (ii) There are 3 marks here so look for *several* ways in which a regional average conceals variations, for example *within* countries as well as *between* countries in the region.

(b) Oman has the lowest GNP (Gross National Product) per capita but note that factors are asked for. So perhaps you should refer also to Oman having the lowest life expectancy?

(c) (i) There are many possible reasons here. If you have studied migration you would think in terms of both the 'push' factors, e.g., unemployment, lower earnings in the countries of origin and the 'pull' factors e.g., shortage of managerial and other skills in the Gulf. Don't content yourself with elaborating on one such point. Migration results from a combination of circumstances and some appreciation of the variety of causes would be expected for the full 4 marks.

(c) (ii) Again there is no single explanation but several reasons are touched upon at the end of the preceding account of migrants in the Gulf States.

(d) Here you will have to reply less on that account and more on a wider knowledge of European migration to areas such as the Gulf.

(e) Now is your chance to make use of what you know about the movement of people into any city you have studied. Note it must be one *named* town or city not a composite account drawn from several cities. Also, you will need to refer to several 'advantages' and several 'disadvantages' if you are to earn the full 6 marks.

Example 8

This question is on a section of the syllabus dealing with 'hostile physical environments' and, more specifically, the distribution and effects of volcanoes and earthquake zones.

(a) **Study the map below.**

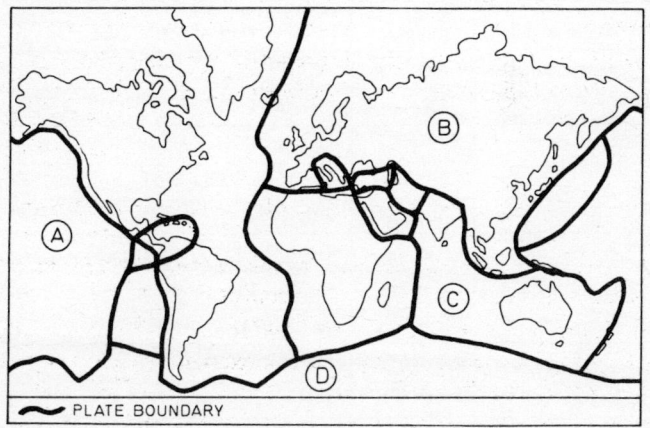

Some smaller tectonic plates have not been named

Name of tectonic plate		
North America	Antarctic	Eurasian
Caribbean	Philippine	Pacific
Nazca	Indo-Australian	Arabian
South America	Africa	

   (i) **Explain the meaning of the term 'tectonic plate'.**   *(2 marks)*

(ii) Name the tectonic plates A, B, C and D. Use the grid below.

(2 marks)

	Name of tectonic plate
A	
B	
C	
D	

(b) **Study the table below**

Earthquakes 1981–1983

Major region	Country	Date of earthquake
North America	USA	May 1983
	USA	October 1983
Central America	Mexico	October 1981, June 1982
	Costa Rica	April 1983, July 1983
	Honduras	July 1982
South America	Chile	October 1981, October 1983
	Colombia	October 1981, March 1983
	Peru	April 1981, April 1983
Pacific Asia	Japan	March 1982, May 1983
	Indonesia	January, February, December 1982
	Philippines	August 1983
Rest of Asia	Afghanistan	January 1982
	India	January 1982
	Iran	June 1981, March 1983
	Pakistan	September 1981
	Turkey	October 1983
Europe	Italy	February 1981, March 1982, October 1982
	Greece	February 1981, December 1981
	Yugoslavia	August 1981, June 1982, July 1983
	Rumania	November 1981

(i) Which major region received most earthquakes between 1981 and 1983? *(1 mark)*

(ii) Describe the relationship between the information shown in the table and that shown on the map in *(a)*. *(1 mark)*

(iii) Explain why earthquakes occur. *(2 marks)*

(c) Study the map below.

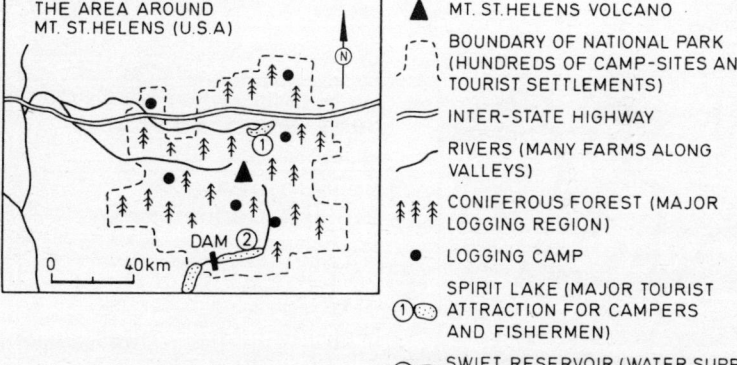

THE AREA AROUND MT. ST.HELENS (U.S.A)

N

0 40km

DAM ②

▲ MT. ST.HELENS VOLCANO

⌐⌐⌐ BOUNDARY OF NATIONAL PARK (HUNDREDS OF CAMP-SITES AND TOURIST SETTLEMENTS)

══ INTER-STATE HIGHWAY

RIVERS (MANY FARMS ALONG VALLEYS)

⚘⚘⚘ CONIFEROUS FOREST (MAJOR LOGGING REGION)

● LOGGING CAMP

①⊂◌ SPIRIT LAKE (MAJOR TOURIST ATTRACTION FOR CAMPERS AND FISHERMEN)

②⊂◌ SWIFT RESERVOIR (WATER SUPPLY, H.E.P. AND FLOOD CONTROL)

(i) Describe the types of natural disaster which might be likely to occur in this area in the event of a volcanic eruption by Mt St Helens.
Give evidence from the map. *(2 marks)*

(ii) Name *two* groups of people in this area who might be reluctant to leave despite warnings of possible volcanic eruption.
In each case give *one* reason why. *(2 marks)*

(d) **Study the data below.**

'California earthquake'

'Atlantis will rise
Sunset Boulevard will fall
Where the beach used to be
Won't be nothing at all
That's the way it appears
They tell me the fault-line
Runs right through here
So that may be.
What's gonna happen
Is gonna happen to me
That's the way it appears'

(1960s pop song)

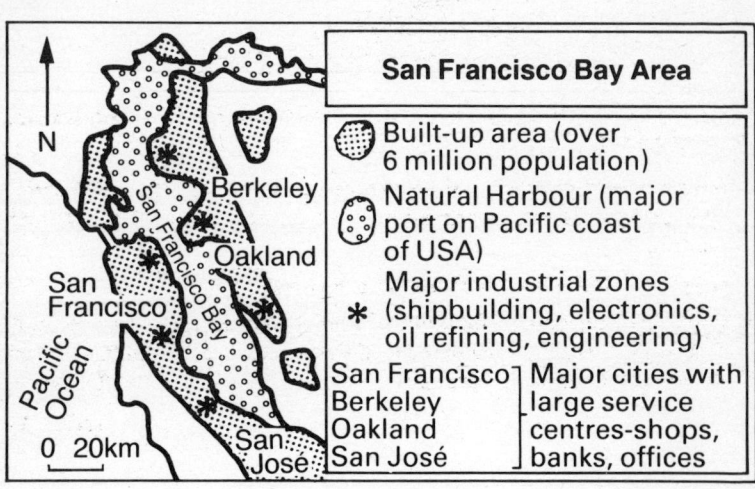

(i) Describe the types of disasters which might follow an earthquake in this area.
Give evidence from the map.   *(2 marks)*

(ii) Name *two* groups of people living in this area who might be reluctant to leave despite warnings of possible earthquakes.
In each case give *one* reason why.   *(2 marks)*

(e) Study the illustration below.

'How to build safer houses'. Government publication (Guatemala, Central America)

Using *named* examples of areas subject to natural hazards, describe the responses taken by the inhabitants to reduce the effects of such hazards.

Types of areas might include the following:

Slopes of volcanoes; plate boundaries (other than those areas in *(c)* and *(d)*), river flood plains; coastal deltas; areas in the path of hurricanes; areas subject to drought. *(7 marks)*

(NEA A)

Even if you have not studied the topic, can you see the kind of answer you would write?

(a) Is a matter of relying on your memory for the definition and identification of the plates marked on the map.

(b) Also mainly makes demands on your memory but don't be content with something as brief as 'movement along plate boundaries' for part **(iii)**. The three lines provided in the answer book and the 2 marks available suggest a little more than that is called for.

(c) **(i)** Is more than just recalling the events when a volcano erupts. Can you *describe* how those events would affect the areas shown on the map? Lava flows of course but think about other possibilities such as a drifting cloud of ash or the rupturing of the Swift reservoir dam. Note 'types of disasters' and 'evidence from the map'.

(c) **(ii)** Just naming two groups of people won't get you very far; a convincing reason why each group might not leave (farmers protecting their property?) is also needed.

(d) Calls for similar abilities of imagining how a different type of disaster, an earthquake, could affect a specific environment, in this instance an urban area.

(e) Again a chance in the last part of a question to use examples of your own to illustrate people's responses to natural hazards. The choice is yours but note examples and they should be *named*. You could refer to several examples or describe two in more detail; either approach would be acceptable given the wording of this question.

With the next three examples, the question is printed, followed by the examiner's provisional mark scheme. Try reading each question in turn and then, without the benefit of guidance on what it means or how it should be answered, anticipate how you would answer it. Then check your interpretation of the question and how to answer it against the examiner's mark scheme.

Example 9

Study the diagram below.

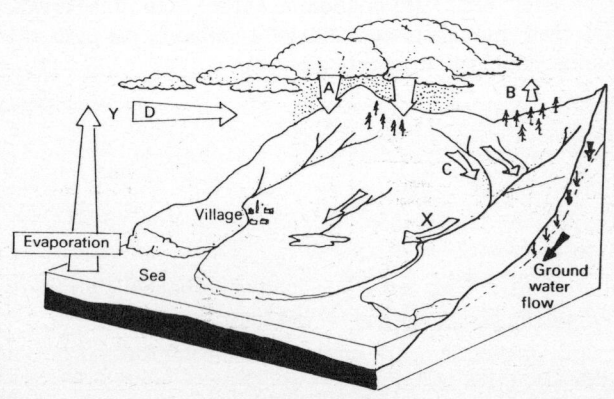

The Water Cycle

(a) (i) Explain why the term 'cycle' is used in the title of the diagram. *(1)*

(ii) From the list of terms in the box choose the correct one for the processes taking place at

A ....

B ....

C ....

D ....

| evaporation |
| transpiration |
| nationalization |
| surface flow |
| precipitation |
| indoctrination |
| interception |
| condensation |
| sedimentation |

(4)

(iii) Explain what is meant by the terms:

'groundwater' 'flow':
'evaporation': *(2)*

(iv) In what form is water being transferred at X and Y? *(2)*

(v) In the hydrological cycle water not only changes form, it is also 'stored' in a number of places. Suggest *four* places on the diagram where water is being stored. *(4)*

(vi) Describe ways in which this hydrological cycle will be altered if the village expanded into a large town. *(4)*

(b) Study the map and graph below giving information about the Severn floods in 1977.

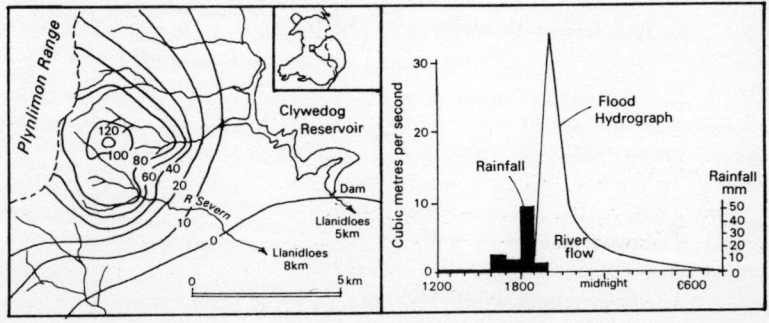

107

(i) What is an 'isohyet'? *(1)*

(ii) Suggest *two* reasons why the flooding in Llanidloes came as a complete surprise to the people living there. *(2)*

(iii) For how long was water flow in the River Severn over 10 cubic metres per second? *(1)*

(iv) What was the delay time between the peaks of rainfall and river flow? *(1)*

(v) How would the graphs be different if the same amount of rain had fallen over a longer period e.g., a day? (You may draw the graphs to illustrate your answer, if you wish.) *(3)*

(vi) By reference to *named* examples you have studied, show how *and why* other rivers may have responded quite differently to the same amount and intensity of rainfall as experienced by the River Severn. *(6)*

(c) For a *named* river you have studied, describe the ways in which river flooding in an area affects the way land is used. Use a sketch map to illustrate your answer. *(5)*

(Total 36 marks)

Marking scheme

(a)	(i)	Because the water is continually being re-cycled	*1*
	(ii)	A Precipitation; B transpiration; C surface flow; D condensation	*4*
	(iii)	movement of water below the surface of the ground when water changes its form from liquid to water vapour (a gas)	*2*
	(iv)	liquid; water vapour	*2*
	(v)	in a lake; in rivers; underground; in the sea; in plants	*4*
	(vi)	Water cycle will be affected in a number of ways; water likely to be polluted; water table drop; more surface storage of water in reservoirs; more rapid run-off; rivers and streams may be put underground	*4*
(b)	(i)	Line joining up points of equal rainfall	*1*
	(ii)	Very little rainfall in Llanidloes itself; the most intense rain was localized over headwaters of Severn	*2*
	(iii)	2½ – 3 hrs	*1*
	(iv)	2 hrs	*1*
	(v)	The rainfall columns would be very low and spread throughout the day. The hydrograph would show a slow steady rise rather than a peak – and a slow decline	*3*

(iv) Should include references to nature of bed rock –
    permeable, impermeable etc.   *(2)*
      vegetation cover   *(2)*
      stream ordering etc.   *(2)*   **6**

(c) Answer could be at any scale and in any context e.g.,
    Nile Valley or a local case study   *1 × 5*   **5 | 36**

(Avery Hill)

Example 10

Contrasts in development

(a) **Study the graphs below which give information on the levels of development of five countries.**

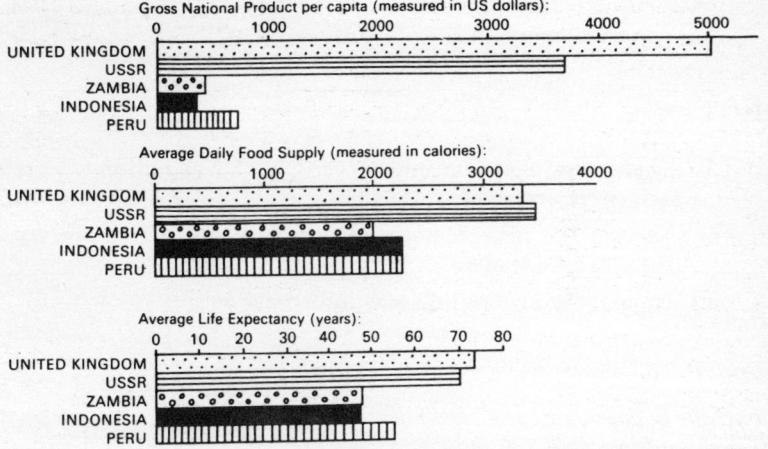

Gross National Product per capita (measured in US dollars):

Average Daily Food Supply (measured in calories):

Average Life Expectancy (years):

(i)   **What is the Gross National Product of the United Kingdom?**
                                                (1 mark)

(ii) **How many calories make up the average daily food supply in Zambia?**   *(1 mark)*

(iii) **Which of the countries shown has the shortest average life expectancy?**   *(1 mark)*

(iv) **Which of the countries shown are part of the developing world?**   *(1 mark)*

(v)   **Use the graphs to state the basic differences shown between the countries in the developing world and those in the developed world.**   *(3 marks)*

(vi) **Give *two* other ways of measuring the level of development in a country and explain why *each* is considered to be a useful indicator of development.**   *(4 marks)* *(b)* Study the sketches below which show how a government land reform scheme has helped with the development of a rural area.

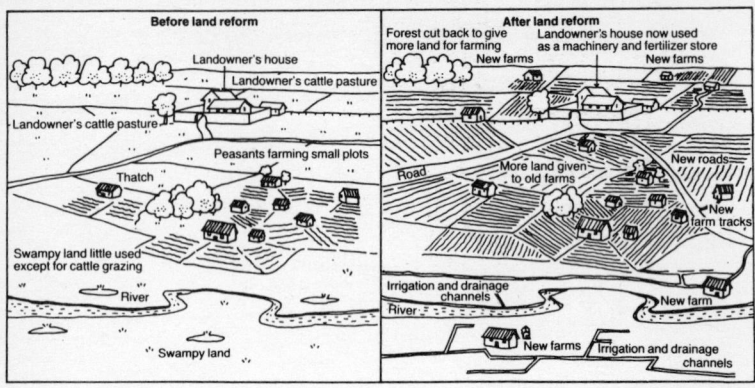

(b) (i) Suggest how a government can carry out a land reform scheme such as the one shown. *(4 marks)*

(ii) Describe the main features of a *named* land reform scheme that you have studied. *(5 marks)*

(iii) Explain why land reform may not always be easy to carry out. *(5 marks)*
(Total 25 marks)

Contrasts in development – Marking scheme

(a)	(i)	5 000–5 050 dollars ($\frac{1}{2}$ without units)	1
	(ii)	2 000	1
	(iii)	Indonesia	1
	(iv)	Zambia, Indonesia and Peru (must have all three)	1
	(v)	Developing countries have lower GNP, less food and shorter life expectancy.	3
	(vi)	E.g. rate of population growth – developed countries have relatively stable populations. Energy consumption – developed countries have greater need for (and can afford) more energy	4
(b)	(i)	Buy the land off the local landowner, or evict him/her as a result of political ideology; allocate new landholdings to peasants; provide money for infrastructure improvements such as the road or drainage/irrigation schemes; provide machinery to help clear and plough new lands etc. Any 4 points.	4

(ii) E.g., a Chinese commune (such as Ping Chow) – *1 mark* – Communist government took land from landowners and handed it over to peasants who were ordered to set up agricultural communes, administered through a people's council. Council organizes which crops are to be grown and how rest of land is to be utilized, how much fertilizer needed, what capital schemes are necessary, etc. *(4 marks)*   5

(iii) Governments may not have the money nor the political support to be able to go ahead with land reform; peasants may not respond or be able to cope with the new demands put upon them; government plans often over-ambitious and optimistic; major landowners bound to oppose plans etc.   5 | 25

(SEG A)

Example 11

(The maps referred to in this question appear on the following two pages.)

Study maps A and B in the Resources Booklet (Fig.5) which show changes in a farming area of upland Britain between 1960 and 1980.

(a) (i) State one change that has taken place in the number of farms.

(ii) State one change in the amount of land being farmed.

(iii) Describe the other changes which have taken place in the area.
(5)

Study maps C and D in the Resources Booklet (Fig. 6) which show changes in a farming area of lowland Britain between 1960 and 1980.

(b) (i) Describe *two* differences in the layout of the fields.

(ii) State *two* other differences which have taken place on the farmland.

(iii) What change has taken place to the actual farm buildings? *(5)*

(c) (i) Explain how changes in farming methods may have encouraged two of the changes mentioned in part *(b)* (i) and *(b)* (ii).

(ii) Explain how conservationists might object to two of the landscape changes that have taken place. *(4)*

(d) (i) Suggest one reason why the pattern of fields has changed in the area shown on Maps C and D but not on Maps A and B.

(ii) Suggest one reason why other economic activities have developed in the area shown on Maps A and B but not on Maps C and D. *(2)*

(e) For one named area outside the British Isles, where changes have been made for economic gain:

(i) describe these changes.

(ii) explain to what extent these changes are an environmental loss.

(9)

(Total 25 marks)

Fig. 5 Farming landscape in upland Britain (For use with Example 11)

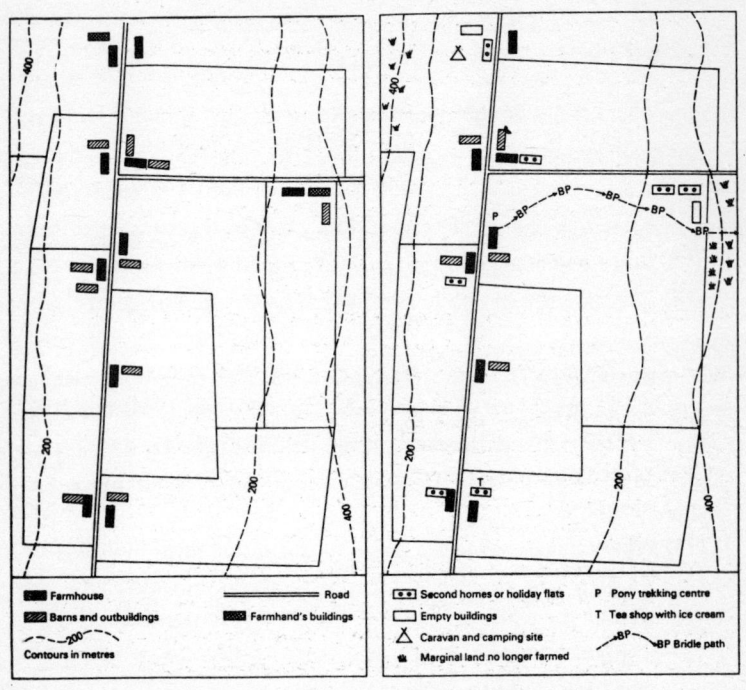

Farmhouse

Barns and outbuildings

Contours in metres

Road

Farmhand's buildings

Second homes or holiday flats   P   Pony trekking centre

Empty buildings   T   Tea shop with ice cream

Caravan and camping site

Marginal land no longer farmed   BP   BP Bridle path

Map A 1960   *Map B 1980*

Fig. 6 Farming landscape in lowland Britain (For use with Example 11)

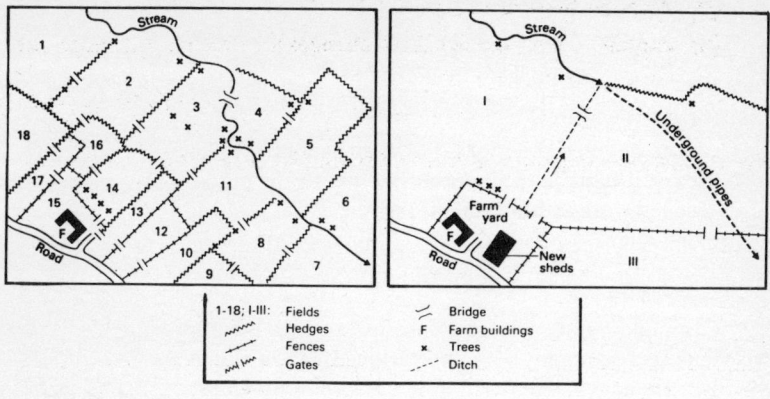

Map C 1960   Map D 1980

		Sub-mark	Total

Marking scheme

(a) (i) **Less in 1980** · 1

(ii) **Less in 1980** · 1

(iii) **Some farms empty: Some farms/barns are second homes: Pony-trekking centre: Tea shops: Caravan site (3 of)** · 3 · 5

(b) (i) **Less fields in 1980: larger fields in 1980: field boundaries are fences not hedges. (2 of)** · 2

(ii) **Most trees have been removed: part of the stream has been piped underground.** · 2

(iii) **Large new sheds have been built.** · 1 · 5

(c) (i) **Large machinery needs large fields: crop rotation is less necessary because of chemical fertilizers etc.** · 2

(ii) **Removal of trees and hedges would create a 'barren landscape'. Removal of hedges and stream would deprive wildlife of their habitat.** · 2 · 4

(d) (i) **Upland Britain is not suitable for large-scale arable farming.** · 1

(ii) **Upland Britain's scenery provides a basis for tourism.** · 1 · 2

(e) **Name area** · 1

113

(e) Level one response
A basic description of the change with only a limited
assessment of environment loss: e.g., the development
of the chosen activity rather than its effects on the
environment.   *2–3*

Level two response
A more detailed description of the changes with some
real assessment as to the resulting loss to the environment, e.g., a link between mining processes and the
environment.   *5–6*

Level three response
A full and detailed factual description with a compre-
hensive assessment as to the various ways in which the
listed changes have resulted in environmental loss, e.g.,
damage to the ecological balance.   *7–8*   8   8
                                                           ─────────
                                                                25

(LEAG B)

Note in this last case that part *(e)* is to be marked in a different way.
There is no list of points to be credited but rather statements about the
quality of the answer expected for a given range of marks. The examiner
would judge which of the three statements best describes the quality of
your answer.

ANSWERING OBJECTIVE QUESTIONS

The decision you have to make when answering objective questions is
which of the various answers offered is the correct one. You will have
between a minute and a minute and a half on each question in which to
make that decision.

The right choice may seem obvious. But remember that each of the
possible answers is likely to be reasonable enough to be worth a look;
examiners don't include so-called 'distractors' which are so ridiculous
that no one in his/her right mind would choose them. So at least take a
second look before confirming your choice by marking the appropriate
letter on the answer sheet.

If you are not sure, work towards a decision by eliminating the
answers you know are *not* correct. If you are still not sure, don't spend
any more time on that question. Before moving on, make a small mark
against the unanswered question on both the question paper and the

answer sheet. Should there be time at the end of the paper you can then quickly locate the questions you have missed out. There may be only a few minutes left to answer the questions you omitted but *make sure you attempt an answer to every question*. Even if you have no idea what the right answer is, bear in mind that a guess has (usually) a one in five chance of earning you a mark.

ANSWERING PROBLEM-SOLVING EXERCISES

Some parts of the problem-solving exercise on page 87 can be answered in the same way as parts of ordinary structured questions. For example, part *(i)* is a matter of reading information from the map provided. But much of it requires you to respond to instructions such as 'give good reasons' and 'explain why'. The last part of the question (not reproduced here) asks you, for 9 marks, to explain how you decided on your choice of route and why you think your route is better. What you have to do is to weigh up all the available evidence and make a reasonable decision. But, more than that, you must be able to explain the reasons which led you to that decision. Only a few marks are for stating the decision; most are for the way you justify it.

A problem-solving exercise tests not only what you have learned but what you are able to do with your geographical skills and understanding. It is the most striking illustration of the point that going into a GCSE Geography exam with your mind full of, perhaps clogged up with, facts will not get you very far.

SUMMARY: NINE POINTS TO REMEMBER

1   *Go into the exam room knowing exactly* what to expect – *the number of questions on the paper, the number of questions you should answer, the approximate time available per answer.*

2   *Organize your* time *by having a schedule of the time you can allow for each question. Don't panic if you are five minutes behind schedule but equally don't leave yourself insufficient time to answer the question you tackle last.*

3   *Start with a few minutes of* careful planning – *choose which questions to answer (if there is a choice), underline all the instructions in those questions* describe, explain, compare, *etc.).*

4   *Don't overlook the* conditions *attached to questions such as 'with the aid of a sketch map' or 'with reference to Diagram 3'.*

5 *Watch for the* number *of examples, reasons, advantages, etc. you are asked to give and make sure you don't mention fewer. If the number is not specified, it's probably safer to refer to several briefly than to dwell on one at some length.*

6 Write legibly. *Experienced though examiners are at deciphering illegible handwriting, they don't have the time (or the patience?) to read through what you have written several times over.*

7 *Don't expect the examiner to understand what you mean if you use your own abbreviations or write in a disconnected series of phrases. As with legibility, you should not expect the examiner to have to work out what you mean.*

8 *Never answer less than the* required number of questions. *Whether you have run out of time or out of ideas, there must be something relevant you could write in answer to that last question. If it's an objective question, your guess could just prove correct. Even one or two marks on the last question could make the difference of a grade in your result.*

9 *If there is time at the end,* read through your answers. *You may notice a mistake or a point you missed first time round.*

SECTION 10

Preparing for the examination

You are probably taking this particular exam, GCSE Geography, for the first time but you will have had more than enough experience during your school career of taking other examinations. If the way you prepared for those exams suits you and has served you well over the years, stick to it. The best way to prepare for exams is a matter of personal choice but here are some suggestions you may find useful.

There are three main stages involved in being ready for the examination:

(i) a good set of notes from your course;
(ii) a clear view of what to revise;
(iii) an effective revision programme.

In short, you need notes to revise *from*, you must be clear what you are revising *for* and know *how* to revise in a way which will work for you.

Each of these stages will be dealt with more fully. Before coming to that, keep one general point in mind – how *little* the examiner can test you on in the time available. All those hours of study and you will have perhaps two hours in which to show what you understand and can do! If one thing is certain, it is that there is no point in even trying to remember everything you have covered during the GCSE Geography course.

NOTES

You will need a set of notes on each part of each topic written in such a way that *you* understand them. A good textbook will help explain and illustrate a point and will be useful to refer back to, but it is no substitute for your own notes. Also, the book's way of dealing with a topic and the depth of treatment may not be appropriate for your syllabus since it is unlikely that the book was written with a particular syllabus in mind. Duplicated notes provided by your teacher are more carefully tailored to the demands of the exam you are preparing for. But they too are written in someone else's words.

If all the exam called for was repeating the chunks of information which you could recall, you would be best advised to try to memorize parts of the textbook or of the duplicated notes. But what should be clear by now is that answering most GCSE Geography exam questions calls for more than a good memory. It requires you to show that you can *use* certain skills, can *understand* key ideas and can *apply* those ideas to evidence you meet for the first time in the examination. Just as mastering skills requires practice as well as instruction, so understanding involves 'processing' ideas as well as listening to, or reading, explanations of them.

The set of notes copied from the book or taken from a duplicated handout may well be clear and accurate, but relying on such notes is not the best way to prepare for an exam. Sooner or later you are going to have to show you can do something *your* way, whether that something is giving a grid reference on a map or explaining the location of a chemical works. Try using your own words sooner and more frequently rather than leaving practice at doing so to the occasional test or exam. The way you are taught the course may give you the chance to write your own notes. Even if that is not so, at least make sure that your revision notes (see pages 119–120) are written in that way.

WHAT TO REVISE?

If trying to remember everything you have studied during the course is unrealistic, what *should* you focus on in your revision? Following the guidance in section 4 will have made you aware of the framework of key ideas and skills around which study of the themes is developed. It is your understanding of those ideas, not the specifics of any one topic, which the examiners will be testing. Remind yourself of that framework of key ideas and skills as you come to revise each theme.

You could try keeping a record of your progress as you go through the GCSE course. Don't judge progress in the traditional way – 'have done' topics X, Y and Z. Check how you are progressing in a more constructive way with a list of 'can do's'. With *skills*, you will 'have done' them all by the end of the course, but so what? You will only be ready for the exam when you 'can do' them. Can you measure distance on a map, read a histogram, draw a sketch map or whatever other skills your syllabus calls for? With *ideas*, you should have covered them all by the end of your course, but do you understand them and can you apply them? If the population geography theme outlined on page 19 was on your syllabus, could you explain 'the causes and effects of variations in the structure and growth of population' both in countries you have studied and in an unfamiliar country, the information about which was presented in an exam question?

If you have not recorded your progress in this way during the course, do so as the first step in your revision of each theme. It will help you use your time most profitably by highlighting which aspects of the course need more time spent on them.

You may also find it helpful to look at past exams, not just question-by-question as suggested in section 8 but across the whole paper. Don't make the mistake of looking at the *topics* of last year's questions and hoping that the same questions will come up again this time. Ignore the topics and ask what can past (or specimen) papers tell you about what you can expect to have to do in an exam?

What type of answer do you have to be able to give? For example, for the question at the beginning of this book (pages 1–3), a list of the demands would look like this:

- extract information from maps, bar graphs (*(a)*) and triangular graphs (*(d)* (i));
- remember definitions (*(d)* (ii));
- understand an account supplied (*(b)* (i));
- name examples (*(b)* (ii), *(c)*, *(d)* (ii));
- remember and explain changes (*(b)* (iii)) and problems (*(b)* (iv));
- understand patterns at different scales (*(c)*);
- discuss a given statement (*(d)* (iii)).

These are the 'can do's' the examination actually gives marks for. Recognizing both what the syllabus says you should be able to do and what past papers expected candidates to do will help you direct your revision. It is not a mass of information which is needed to prepare you for answering that question on unemployment and industry. It is a modest amount of information and understanding plus the ability to use the skills you have acquired to respond appropriately to the particular demands which the question makes.

PLANNING YOUR REVISION

How do you usually set about revising for exams? A quick skim through all the notes you have in the hope that something might stick in your memory? Or a careful line-by-line going over the notes with the probable result that revision time runs out with half the notes still unread? Try to be more systematic even if, as with all well-laid plans, you eventually find yourself falling some way short of your best intentions. Your plan should include a method of taking *revision notes* and a *timetable* for your revision.

REVISION NOTES

To reinforce your *understanding*, make revision notes as you go through, summarizing the main general points on each theme. For example, on migration, you might list common 'push' and 'pull' factors. On the drainage basin (page 107) you could list the main inputs, throughputs and outputs. However, with a topic such as that, it may be more useful to make revision notes in diagram form. Many people find it easier to remind themselves of the connections between points on a diagram than to recall a set of notes.

Before committing details of the *examples* you have studied to memory, ask yourself how many examples on each topic and how much on each example you will need to refer to in an answer. Sometimes you are are asked to do no more than name an example. On the other hand, substantial parts of some questions (pages 74 and 77) invite you to write at some length about an example you have studied. What is certain is that you will have studied many more examples than you will need to refer to in an exam and many examples will have been studied at greater depth than you could make use of. If you are to economize on time in your revision, do it in the number of examples you try to remember and how much you try to recall about each.

The only sensible way to revise *skills* is to practise them. Don't just look at your notes on, say, climate graphs; pick out some examples from a textbook or atlas and see if you can interpret them.

A REVISION TIMETABLE

The shape of your timetable will depend on many things, including how many other exams you are taking and how much time you are willing to give to revision. Don't set yourself too taxing a schedule. Even if your determination never wavers, there are bound to be interruptions and some things will take longer to revise than you anticipate. When planning your timetable think about these points:

1 Allow for a thorough revision of all your course notes in the weeks before the exam, during which period you will make your revision notes. But leave time also for a last-minute refresher, perhaps the night before the exam, when you will rely wholly on those revision notes to jog your memory.

2 Don't allocate your geography time equally across all the themes and the skills. Your record of progress, whether taken during the course or at the start of revision time, will have highlighted strengths and weaknesses. You should allocate revision time to where it is most needed.

3 The revision process can sometimes seem endless. Give yourself the encouragement of marking off on a list the themes and skills you have revised.

4 Within the time you have allowed, make provision for testing yourself as well as for making revision notes. The testing could consist of doing something like drawing a sketch map from memory or it could involve enlisting the help of a friend or parent to ask you questions.

5 Within each block of revision time, will you concentrate only on one subject or would two subjects per evening, perhaps starting with the subject you find less interesting, be better?

6 Within that block of time, will you plan for a series of short bursts of revision punctuated by several breaks? Or would a sustained effort be preferable with some relaxation - perhaps a favourite TV programme - to look forward to when you have finished? Short bursts may seem attractive, but you would have to make the effort to concentrate again each time you go back to the revision. The sustained effort means you only have to settle down to it once, but for how long can you maintain your concentration?

7 At the start of each revision session you could begin by writing down the main points you can remember on that topic. Referring back to it when you finish the topic will show what you have gained from the revision and help boost your confidence.

What this book has tried to give you is the kind of confidence which can come from understanding the demands which the GCSE examination will make. You will obviously develop your geographical abilities during the course but you should also know how to use your abilities in ways which will be recognized in an examination. Enjoy your study of geography and good luck in your examination.